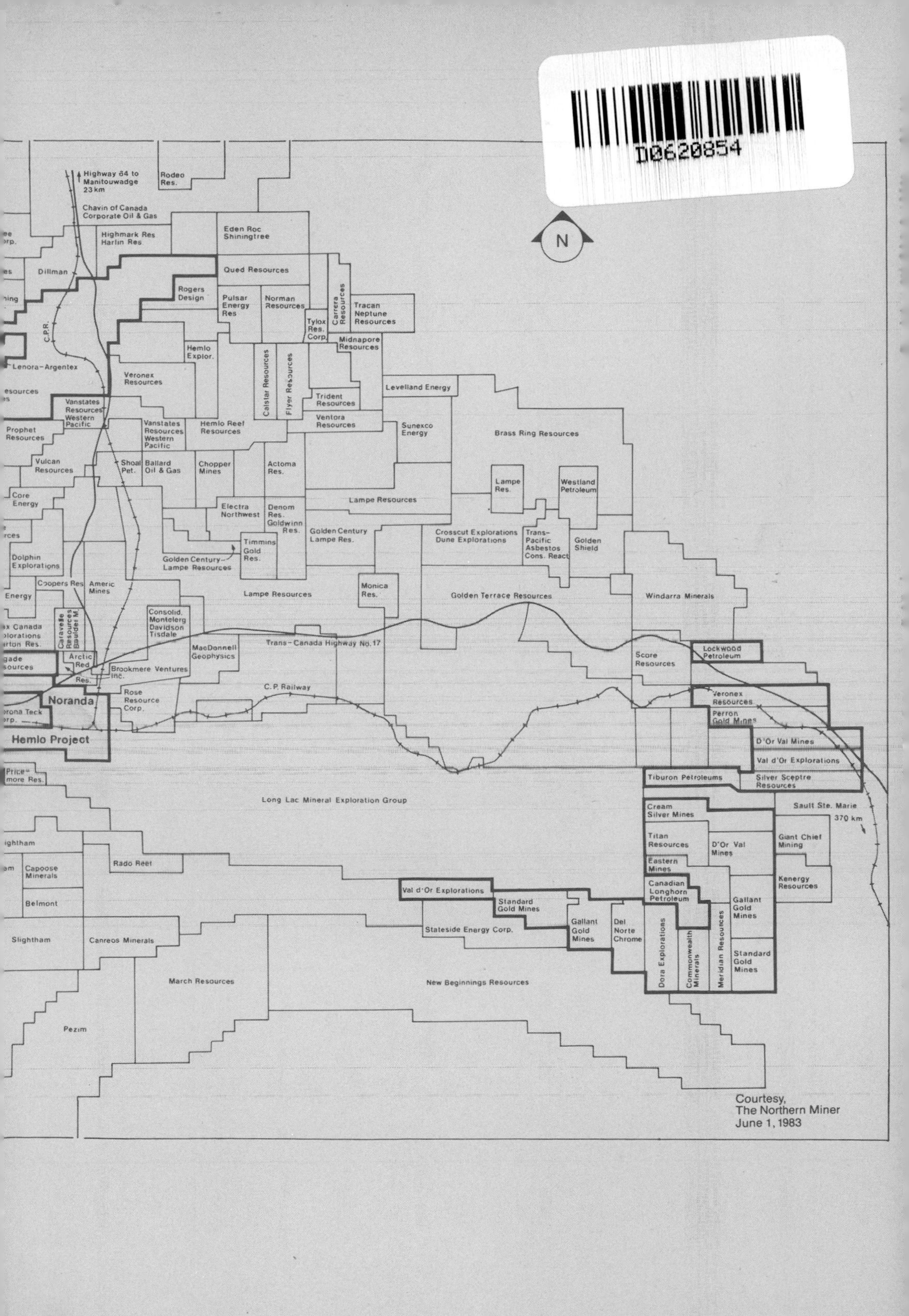

Courtesy,
The Northern Miner
June 1, 1983

GOLDEN GIANT

GOLDEN GIANT

Hemlo and the Rush for Canada's Gold

MATTHEW HART

Douglas & McIntyre
Vancouver / Toronto

Copyright © 1985 by Matthew Hart

85 86 87 88 89 5 4 3 2 1

All rights reserved. No part of this book may be reproduced or transmitted in any form by any means without permission in writing from the publisher, except by a reviewer, who may quote brief passages in a review.

Douglas & McIntyre Ltd., 1615 Venables St.
Vancouver, British Columbia V5L 2H1

Canadian Cataloguing in Publication Data

Hart, Matthew, 1945–
Golden giant
ISBN 0-88894-467-5

1. Gold mines and mining – Ontario – Hemlo.
I. Title.
HD9536.C23H45 1985 338.2'741'0971312
C85-091056-0

Typeset by Alphatext
Printed and bound in Canada by Imprimerie Gagnée Ltée.

This book is for my wife,
Sylvia Alden Morley

Contents

The Hemlo Timetable

2,000,000,000 B.C.	Submarine volcano erupts, spilling gold-bearing brine into the region.
1,000,000,000 B.C.	Geologic structures shift and the gold thrusts up closer to the surface, in some places breaking through.
1869	Moses Pe-Kong-Gay, an Indian prospector, locates gold showings near the present town of Heron Bay.
1872	Shafts are sunk on Pe-Kong-Gay's discovery, and a little ore is shipped.
1920	J. LeCour, CPR station agent at Hemlo, sinks a few test pits near the area of the present strike. Property not developed.
1931	Ontario Department of Mines recommends the area for further exploration.
1944	Peter Moses, an Indian from Heron Bay, discovers gold in the present strike zone. A minor staking rush ensues.
1945	Dr. J. K. Williams of Maryland stakes the eleven claims which have remained in his family to the present day, and which are in the heart of the present strike zone.
1947–65	Various companies explore the area, drilling, trenching, and defining up to ninety thousand tons of ore: insufficient for development. Properties abandoned.
1965–78	Hemlo properties staked, restaked and finally lapse for want of work.
1979	In December, the Hemlo claims come open.

Prospectors John Larche and Don McKinnon pounce, staking the first claims.

1980 Larche and McKinnon take their first fourteen claims to Bay Street and attempt to attract backers. Repeatedly they fail. Finally promoter Don Moore helps them strike a deal with Steve Snelgrove, who promises to arrange Toronto financing. Bay Street financiers Claude Bonhomme and Rocco Schiralli grubstake Larche and McKinnon, who stake, first, 156 new claims and, later, 12,000 more. Snelgrove fails to strike a deal in Toronto for the original claims, and with the permission of McKinnon and Larche he takes the property to Vancouver and pitches it to Murray Pezim and his associate, Nell Dragovan at Corona Resources. Corona options the property.

1981 Dick Hughes and Frank Lang of Vancouver option the 156 claims held by Larche, McKinnon, Bonhomme and Schiralli. Commissioned by Corona, Timmins geologist Dave Bell begins to drill the company's fourteen claims. Initial results are disappointing. At Hole 76, Bell catches the motherlode. Corona stock goes from $1.29 to $34 a share. In this year Noranda Inc. buys control of MacMillan Bloedel and Brascan's Edgar and Peter Bronfman buy control of Noranda. Soon after, the entire natural resources industry slumps worldwide. Noranda hits hard times.

1982 Toronto mining magnate Peter Allen's Lac Minerals strikes gold on Dr. Williams's claims. The Rothschilds purchase thirty thousand shares of Hughes's and Lang's two Hemlo companies, Goliath Gold Mines and Golden Sceptre Resources. Goliath–Golden Sceptre strike The Golden Giant, the site of the richest ore at Hemlo. Noranda moves onto The Golden Giant deposit, optioning the Goliath–Golden Sceptre property. This move by mighty Noranda sparks a staking rush. In this year, Teck Corp. options the Corona claims. Teck-Corona, Noranda and Lac Minerals become the three major players on the site.

1983 On 17 January the Vancouver Stock Exchange trades 33.3 million shares, a Canadian record. Most of the activity is in Hemlo stocks.

1984 Murray Pezim loses control of his Hemlo property. In addition, a major portion of the Hemlo site becomes a matter of disputed ownership in a case to appear before the Supreme Court of Ontario in October 1985.

1985 In January Noranda began to bring up ore and test its mill, and by April was to have cast the first bars of gold, beating its two competitors into production. The Hemlo gives up its treasure at last.

Acknowledgements

My greatest debt in researching this book is to those key Hemlo players who gave so freely of their time and knowledge. John Larche and Don McKinnon — who staked The Hemlo — are a credit to that race of canny men who sniff the northern rock for treasure. The candour, consideration for my ignorance, and endlessly intriguing anecdote of the two prospectors made The Hemlo staking come alive. Murray Pezim is a man of great courtesy, at least to puzzled writers, and watching him perform amidst the turbulence of an ordinary (for him) day is an experience I will not soon forget. To Frank Lang, thanks for sharing your admirable passion for what you do so well: develop gold mines. Nell Dragovan is a busy woman, and that she gave me the time she did speaks highly for her. Dave Bell, the Timmins geologist whose theory of the volcanic origins of the gold led to the motherlode; the respected investment dealer and goldbug Joseph Pope of Pope & Company; Alf Powis, chairman of Noranda Inc.; Peter Allen, president of Lac Minerals Ltd.; geologist Fenton Scott; Mort Brown, editor of *The Northern Miner;* Bob Brearley and Rob McEwen, both of McEwen Easson Limited, stockbrokers; Michael O'Shea, Merit Investment Corporation; Ian Hamilton, general counsel, Lac Minerals; T. P. Mohide, Government of Ontario; Barry Sparks, vice-president of Royex and president of Sparks Financial Associates — each of these deserves a pageful of gratitude for his patience and his careful words of explanation. To all who helped go my heartfelt thanks and lasting appreciation. Any blunders are entirely my own.

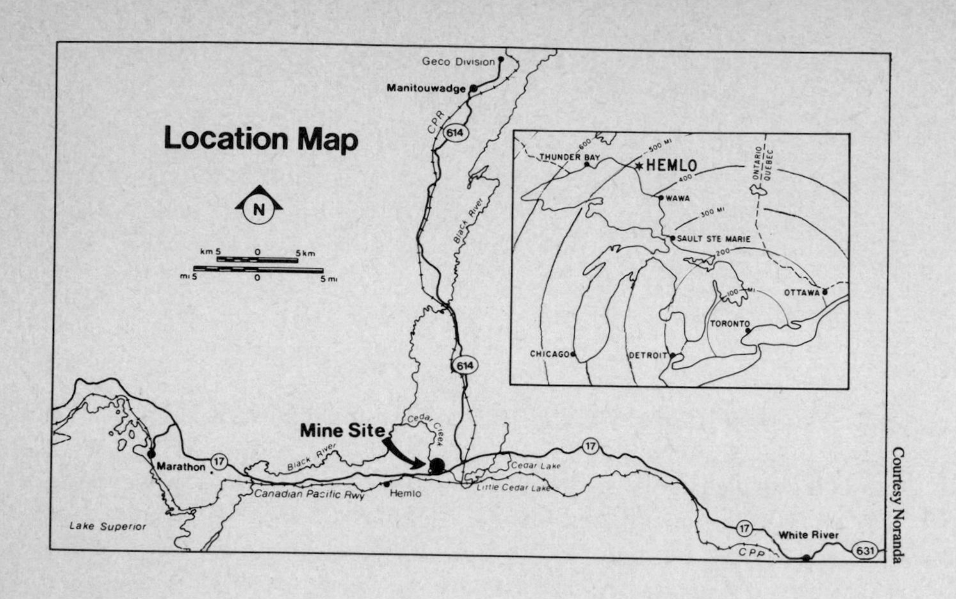

Courtesy Noranda

Prologue

Ever since the first curious human reached into the clear waters of a rushing stream and plucked out a glittering nugget, discovering that it could easily be shaped and stored and that it never lost its lustre, gold has been money. Gold is a metaphor for wealth and power. That is why kings wear it upon their heads. Sceptres are made of gold. Men kill for gold. The very mention of gold can send ten thousand men scrambling across a continent to die in snow-swept mountain passes in their rush for wealth. Armies march to the clink of golden coins. Gold rules the small dark place in the human brain where greed is born and lives.

That is why men look for gold.

And that is why there is a place on the Trans-Canada Highway in northwestern Ontario where three tall headframes rear from the Pre-Cambrian rock. They mark the place where three giant companies scramble for gold. A couple of years ago there was nothing there, nothing but a blank wall of bush. Nothing but a blank wall of bush and the richest deposit of gold-bearing ore in the Americas. It sat there for so, so long, a great slab of fabulous rock just beneath the surface, tilting at an angle, plunging away steeply to the north. For so many years people drove past that place, looking at nothing but the sign that told them Marathon was another forty kilometres west, or White River forty east.

And then the ore was found. Time after time the gold-seekers punched their diamond drills down into the rock, every time pulling up another core loaded with the gold-bearing ore, stunned at the magnitude of their discovery, growing feverish as each day extended the size of the deposit. Even now, no one knows how big the ore body is. It drops away sharply to the north, drops away elusively past the reach of the probing diamond drills. But no one really cares. For there is enough ore in that ground along the highway to keep them mining for twenty years. Plenty of time to look later; plenty of time to track that deposit where it tumbles off down into the granite fastnesses of the geologic past.

That is why if you drive past that place on the Trans-Canada Highway now, just west of the road that leads north to Manitouwadge you will see a frenzy of labour. Three mining consortia — Noranda, Teck-Corona and Lac Minerals — are pouring a combined investment of $750 million into that rocky wasteland, driving down foot after desperate foot in a rush for the ore. By the time you read this, the first of the gold should be out of the ground and cast into glittering bars.

There are always buyers for gold. They are cautious people, people who have learned the lessons of war and treachery, people who distrust paper money and believe that only gold is cash. Some of them live in Switzerland, and some of them live in Russia. Whoever they are, they have heard of that place in the Canadian bush where corporations tear at the rock with their fierce machines, with their dynamite and glycerine and greed. In Moscow and Zurich and London and Johannesburg they know the name of that place.

Hemlo.

The name burns with the lustre of wealth. "Hemlo" is the sound of money, calling from the rugged northern bush to be claimed by the men with the wit and the cunning to find it. Hemlo is the dream the searchers dream when they make their camps among the granite bluffs, tired after the day, anxious for tomorrow.

But a gold deposit does not inhabit a world all its own. The allure of a gold strike is burnished or dulled by the market forces that operate in the great financial centres of the world. Supply and demand, war and peace, truth, rumour, lies: all of these swirl about the globe, and the crafty traders who manipulate the vast pool of bullion listen, make their decisions and plunge. Sometimes at the end of the day they are very rich men. Sometimes at the end of the day they are borrowing bus fare. But they are all back next morning. Gold does not let you go.

Such is the fascination of gold that even when it is tumbling on the world markets it still commands headlines. On the morning of 8 January 1985, four very canny men filed into the London premises of the venerable international banking institution N. M. Rothschild and Sons Limited. Just inside the front entrance, these gentlemen turned off and trooped into a small, oval, elegant room and took their places at a long table. A fifth man, representing the Rothschilds, took his place at the head of the table, and the morning's business commenced.

The five men present at the table represented the principal London bullion dealers, and they were meeting for the first of the two daily "fixings." The London fixings are the benchmark prices for gold traders around the world, and they are awaited anxiously. The bankers met at 10 A.M. — 5 A.M. in New York and Toronto. By six o'clock their decision was leading early-morning newscasts throughout eastern North America: gold had fallen below U. S.$300 an ounce. When trading ended in London that day, the price was $296.50 an ounce. By this time, an aggressive defence of the metal had been launched in New York, and when the market closed on Wall Street five hours later, traders had bid the price back up to $302.50.

But no one really expected the line to hold, and it didn't. Gold sank again, dragging the spirits of its defenders down as it slid beneath the psychologically important horizon of $300 an ounce. Surveying the gloomy scene at the time, Peter Hug, vice-president and chief trader of Guardian Trust, a major bullion dealer, said, "I don't think the downward trend has finished. It's quite possible we could see gold fall to $250 an ounce within the next six to nine months."

At one Hemlo mine alone, the owners estimate that they will have spent almost $300 million before they cast a single brick of gold. And so the bankers tremble, watching the price of gold descend. The venture capitalists, the bright-eyed bold promoters, the risk-takers — all of these watch the news from London and Toronto and New York. No one speaks it aloud, but each is asking the same question: Will gold fall far enough to kill The Golden Giant?

Overwhelmingly, the answer is "No."

In a story carried by its member papers in January 1985, the Canadian Press quoted a financial analyst who dismissed as unfounded any fears that the mines at Hemlo were threatened by the dropping gold price. Said he: "For them to close, the price would have to be two-thirds lower than it is now. The odds of that happening are the same as the U. S. government paying off its debt and the Toronto Maple Leafs winning the Stanley Cup."

Ian McAvity agrees. McAvity is an animated, Toronto-based financial pundit who has published his successful newsletter, *Deliberations,* for thirteen years. *Deliberations* has subscribers in forty-two countries, and in it McAvity deals out the kind of advice on the money market that people are prepared to pay for. Basically, McAvity's message is that as long as the U. S. dollar continues to dominate the world's currencies, people will choose to hold dollars instead of gold. "But eventually the U. S. dollar has to break. As long as the dollar goes through the roof, it understates inflation. Once the dollar breaks

— and I think it will happen soon — then gold regains its role as a savings currency. I believe gold will be over $500 an ounce by the end of 1985, and it could be a lot higher.''

Gordon Bub is an analyst for the Toronto brokerage firm of Gardiner Watson Limited. Bub has been watching the precious metals industry for seventeen years, and he sees nothing but a glowing future for the Hemlo mines. ''First of all there's the cost. They can take gold out of there for $150 to $160 an ounce. Compare that to other mines in this country, where their cost is as high as $250 an ounce.

''Then there's the *size*. Usually a new gold mine outlines production for five years. At Hemlo they've got twenty. Hemlo is just so massive, there's so *much* gold per vertical foot. By the time all of those mines are going flat out in 1989 — when they're all built up to top milling capacity — they'll be producing about 750,000 ounces a year. That's twenty-three metric tonnes of gold a year. If Hemlo were a single mine, it would be one of the biggest in total production in the world.''

The largest gold mines anywhere in the world are in South Africa's fabulous Rand fields and in the Soviet Union. Joined to the rivers of gold pouring steadily from the great mines of these two mighty producers has been the surprising burst of production from Brazil, where alluvial deposits in the interior have attracted pell-mell a swarm of gold-scrabbling miners in a rush reminiscent of the glory days of California and the Yukon. But Gordon Bub will stick with The Hemlo.

''In South Africa they're not putting in any new mines, and the grades [richness of the ore] are getting lower. And of course there's the political situation. Some European institutions won't even touch South African mining stocks.

''As for Brazil, it's difficult to operate down there. You have to have political connections. It's hard to get drills in because of the foreign currency controls. So it's very difficult to get people to buy a Brazilian gold stock. A Swiss banker wouldn't go near it. But he will buy a Canadian stock.''

Oh yes, he will. He has. According to Bub, there has been a lot of European activity in the Hemlo stocks. Certainly the British have been interested in the new goldfield right from the beginning, with Rothschilds moving in to grab a large stake in one of the first stock issues before it even hit the floor of a Canadian exchange. These are not gamblers, nor are they hasty people. They have not bought Hemlo on a hunch. They have bought it because they know better than a lot of people who live forty miles down the road just how many ounces of gold are likely to come out of that bush-covered rock. And they have made their investments, these people from the old, careful banks, not because they expect the price of gold to drop out of sight. No indeed. They expect it to rise. And they are not alone.

The Ontario Legislature is housed in a fanciful tumble of red stone Victoriana set into the middle of Queen's Park in midtown Toronto. Except for a rather absurd equestrian statue of King Edward VII, the northern half of the park is an island of tall, graceful trees, in summertime the haunt of brick-red, gasping joggers and throngs of lunching bureaucrats. Fat, glossy squirrels and stout, teetering pigeons compete for the whole-wheat bread and carrot cake that daily sprinkle the lawns of the old park. There would probably be even more wildlife shuttling about on the grass were it not for a pair of peregrine falcons lodged atop the Whitney Block, which houses the Ministry of Natural Resources. The peregrines were installed on the roof of the Whitney Block by ministry biologists seeking to reintroduce the predators to their range. They have thrived. But they are by no means the only keen-eyed fauna to call the venerable government building home.

Two old cannon guard the west entrance of the Whitney Block, pointing directly across Queen's Park Circle at the Legislature. Behind the cannon dwells Thomas Patrick Mohide, director of the Mineral Resources Branch.

No one was worried about economies of space when the Whitney Block was erected. Its halls are high and wide. Sunlight spills in everywhere through the big, noble windows, and compared to the warrens where most of the civil service work, the Whitney is the Waldorf. Tucked into the northwest corner of the fourth floor at treetop level, with tall windows on two sides and the gothic peaks of the University of Toronto visible in the distance, with a nice big sofa and a great big desk, is the office of Mr. Mohide. Mr. Gold.

Mohide is a tall man, with grey hair slicked straight back from his forehead, as if to clear a line of fire. He beholds the world through a pair of appraising blue eyes. This day he is boldly attired in a pink-and-white striped shirt. His black trousers are held in place by suspenders. His tie — blue and red diagonal stripes — looks as if it would be at home in the mess of a Guards' regiment. It would. Mohide was an officer in the Guards' armoured division during the Second World War. But it was after demobilization that the real passion of his life began.

Mohide started as a trader on the London Metals Exchange, dealing principally in copper and silver. From London he moved to North America, working in the precious-metals divisions of several of the mightier multinational mining companies before coming to rest at Engelhard Industries in New York City. There he became a bullion trader, dealing in gold, silver and palladium in one of the most vigorous and wily marketplaces in the world. Engelhard is the largest purchaser of gold in the United States, so Mohide was not exactly without opportunities to master the subject. "My role was simply to make money on the bullion side, buying or selling advantageously."

Simple? Sure. And General Eisenhower's job was simply to win the war.

In 1972 Mohide was seduced away from Engelhard by an offer to assume the presidency of the Winnipeg Commodity Exchange. He did not exactly linger in the post — he stayed only two years — but while he was in Winnipeg he managed to introduce an innovation that has since then become an integral part of world gold dealing. He created the world's first futures market for gold.

This is the man who now inhabits the large, cluttered office on the fourth floor of the Whitney Block, playing the gnome to his political masters across the street. A lot of the clutter has come from his own busy pen — thick, authoritative analyses stamped with Mohide's imprimatur as director of the Mineral Resources Branch. One of these documents is visible evidence of Mohide's qualifications to speak on the matter at hand. It is a 288-page publication known as Mineral Policy Background Paper No. 12, and it is titled: *GOLD*.

"My function here is not to call the daily fluctuations of the bullion market, to say where gold will go in a week or a month. And I am not here to be a gold booster. I am here to say what I think will happen in the long term. And I do."

And what he thinks will happen in the long term is music to the straining ears of the investors of Hemlo, gathered in early 1985 around the ticker tape like mourners at a wake. For what Thomas Patrick Mohide thinks and says is that gold is going up. Take this business of the U. S. dollar, for instance.

"The Federal Reserve in the United States is printing out money like crazy. They are just printing paper. There's nothing *behind* it. Usually this causes inflation. Maybe not right now, but eventually. Just compare the U. S. dollar to a 1940 dollar. They are exactly the same in every respect. But compared to 1940 buying power, a dollar today will buy you four and a half cents' worth of goods.

"We have a name in the bullion world for what they're printing down there. We call if funk money."

Mohide ponders the profligacy of the Federal Reserve. "The United States is seen as strong and safe. The super-printing of dollars has not dented that optimism yet. But it will. If you print money with nothing fresh behind it, you're diluting the existing currency. It's like putting sand in the bearings of an automobile. It won't show up for a few miles, but when it does show up you just stop."

But the major long-term factor involved in driving up the price of gold, Mohide believes, is the anticipated reduction in output by the unparalleled giant of the market, South Africa. Increased costs and declining grades have forced the South Africans to admit that major drops in production are coming before the end of the 1980s. The market reaction to this is simple: less gold means more money for what gold there is.

"By 1986 the South African decline in production will begin to make itself felt on the gold price. Between 1987 and 1990, South African production is expected to drop by about twenty-five tons a year. And these are South African figures. They come from their own Chamber of Mines."

So concerned are the South Africans about the state of their mines that Harry Oppenheimer, one of the richest mining tycoons in the world, recently lumped six of his mines in the Orange Free State into one single company in order to hold down costs. According to *The Financial Times* of London, the mines involved are all past their prime, and though there are still large reserves of gold, the grade is poor.

To demonstrate the enormous amount of gold production that will be withdrawn from the market by the South African decline, two researchers from the U. S. Bureau of Mines in Denver, writing in the journal *Mining Engineering,* use Hemlo's Golden Giant as an example.

> This district represents the largest development in Canada in a few decades. *It is one of the most significant ongoing developments in the world gold industry.* Noranda's Golden Giant mine is scheduled to reach full capacity in 1987 at 291,000 ounces a year. But to compensate for the [South African] decline outlined above would require developing at least *26 new mines on the scale of The Golden Giant during the next 25 years,* while maintaining the current high levels of annual output in the other major producing nations.

The conclusion is that the world gold supply is headed for decline, and the amount of new gold slated for production cannot even begin to replace it. There are people for whom this is good news. The Rothschilds, for example, will be glad to know it. For one of the richest little gold factories on the planet is about to chug into operation, and they have a piece of it.

Hemlo. A name that used to mark a little, trackless patch of bush between the Trans-Canada Highway and the Canadian Pacific Railway's transcontinental tracks. If you had asked folks fifty miles down the highway about it, they would not have known what you were talking about. But now they do. Hemlo. They know the name in Johannesburg and Rio, London and Zurich and New York. They all know exactly where Hemlo is. To the inch.

And they know more, those men who inhabit the clever chambers of the Credit Suisse and the Moscow Narodny Bank. They know the names of the great corporations at work there, the names of the corporations' leaders and probably the names of the leaders' secretaries' lovers. But what they may not know, and ought to, is that in Canada it is not these watchful titans of finance who *find* the gold. It is just men.

Northern men.

One

The Searchers

Terminal 2 at Toronto's Lester B. Pearson International Airport was designed by a man for whom walking represented the highest social good. There is so much distance between the entrance and some of the terminal's gates that Air Canada operates electrically powered trains to cart you around. These resemble baggage trolleys and are piloted by blank-faced, cheerless Air Canada employees, people who look as if, had they their choice, they would pack you out to the highway and let you hitchhike.

But if you are going north to Timmins, Ontario, get ready for a walk. The Timmins plane is loading at Gate 71, the farthest gate from the terminal entrance. The only way Air Canada could get the plane any farther away would be to park it in Guelph. There is a northern flavour in the cabin as soon as you enter. Two or three seats hold Indians returning home after short visits to the city. They appear to be holding their breath until they can take a gulp of something that will not immediately start to rot their lungs. A few of the other seats hold off-the-rack Conrad Blacks, middle-level corporate maintenance men running up for a few days to impress the secretaries in Sudbury and North Bay. They are studying the *Report on Business* so closely that it might contain instructions for flying the plane. The stewardesses cluster about these pin-striped beacons as if they were the only trees in a desert.

The rest of the plane is filled with all the regular traffic of a northern shuttle. There are men in plaid shirts and jeans, with work boots or sneakers or old, scuffed loafers. There are people for whom the only concession to a trip south is a soiled, lime-green leisure suit of such perfect monstrosity that it might be a careful insult.

A lot of the people aboard know one another, for a northern flight is just the bus home. When people file into the cabin and make their way down the aisle loaded with booty from Eaton's or Simpson's or Bargain Harold's, they peer around to see if they have a friend aboard. They usually have. To the people of the north, a plane is only another jeep.

There is one stop on the flight to Timmins: North Bay. The plane comes in low over the wild chop of Lake Nipissing, banks and lines up for the runway, dropping fast. This is no ordinary northern strip. There are rows of jet fighters lined up along the tarmac, their wings folded up and back like perched hawks. The DC-9 squeaks down, howls into reverse thrust and taxis past the nose of a Hercules C-130, whose turboprops blur the air as the crew waits for the liner to pass. This is the northern command of NORAD — the North American Air Defence System — the nuclear umbrella run from inside a mountain in Colorado Springs. As the DC-9 whines out of the military sector towards the civilian terminal, a white passenger jet with U. S. Air Force markings glides past. A little blue flag flies from the tiny mast just in front of the cockpit window, and on the flag are three gold stars. The brass is up for a look-see.

On the ramp at North Bay, we wait. Ordinarily planes pull in on these northern flights, toss out their passengers and baggage with no more care than the Post Office, and are off again before the mosquitoes have a chance to board. But today we are waiting for some regional carrier to arrive. There are people aboard trying to make the connection for Timmins, and it would be unneighbourly to abandon them. A few passengers stretch their legs at the base of the gangway, chattering with the pilot who peers from the open cockpit window. Fifteen minutes late, a Twin Otter drops down out of the overcast and bounces across the apron. Two or three ragged figures emerge; one heads for the DC-9. One.

This is the north.

The airport in Timmins is jammed. Miners are filing out to board the Austin Airways twin-engine propeller plane that is coughing and snorting away on the runway. It is a new plane, but it looks like an old Viscount, fat and reliable and brave. This is the shuttle for miners at the Detour Lake gold mine. There is a road into Detour Lake, one hundred miles away in the bush, but it is so treacherous and killing a road that the company uses it only for trucking heavy machinery. And gold. The miners fly in for a week at a time and work a straight seven days, twelve hours a day. Then they fly out for a week to rest or drink or kiss their kids, and fly back in. The mine operates twenty-four hours a day, three shifts. The traffic is constant. In and out the miners fly, crowded into the droning yellow aircraft of Austin Airways.

Along one wall of the passenger lobby are the airlines' counters: Air Canada and Austin. Along the other are the rental car counters. There are more rental car companies in Timmins airport than there are airlines. In the north country you do not hop the express into town and buy a subway pass. You rent a car or you walk.

We are here to see two northern men, John Larche and Don McKinnon, prospectors. In the north prospecting is not a particularly exotic profession. If you are good at it, then you will lead a comfortable, middle-class life, buy your kids new parkas every couple of years, take them to Toronto once in a while to shop. Striking it rich is the *idea,* sure, but rarely the *reality*.

This time it is the reality.

The men we have come to see have struck it rich, all right. They have struck it so fabulously rich that it is difficult to conceive of the wealth they have and will get. Their stock alone has made each man a multimillionaire. That is just for starters. No big deal. There are any number of mere lawyers and doctors tooling around Toronto in BMWs who are millionaires. A millionaire is just a middle-class guy with stickier fingers.

These are the men who staked Hemlo, a three-hundred-mile drive west of Timmins. Their *real* money starts to come with the gold. Then the wealth is staggering: $75 million? $100 million? The head swims with these figures. They mean so much that they mean nothing. You cannot translate $75 million into a house or two cars or a private jet. You can subtract *things* from $75 million forever, it seems: the interest will just keep filling in the holes.

To get to John Larche's house, you turn off the main road into town from the airport and take a left at the first traffic light. This is suburbia. If there is a middle-class ideal grafted onto this strike-it-rich city of gold miners, this is it. The streets are neat and the houses for the most part scrupulously maintained. Lawns are laid out, hedges flourish, trees are planted. The only suburban markers absent from this tidy network of streets are tall, leafy, shade trees. But this is the north, where trees take a fearful time to grow. There is only one thing, in fact, that grows faster up here than it does in the south. Fortunes.

Churchill Avenue does not look like the kind of street where a man with $75 million would choose to live. It is a street of bungalows and splits that peters out into a field at the end. Across the field, never very far from anything, the spiky jack pines crowd off into the boreal forest.

Larche's house is a large split-level with a two-car garage. It is painted white and dark red. There is a little clump of birch on the front lawn. Two tubs full of marigolds flank the path leading to the front door. There is no Rolls-Royce parked in the driveway. You would be a lot more likely to see a muskeg tractor with a backhoe mounted on it than you would a Rolls, anyway. The backhoe is at work, in the capable hands of Larche's son David, a full-fledged mining contractor and prospector in his own right.

It is impossible not to like John Larche straight off. He holds himself like a man whose life has been reduced to a few elements of simplicity, some wisdom that he holds right there with him at all times. It means that he does not have to dissemble, that he does not try to be anything but exactly what he is, a good man who loves his kids and knows his way around in the bush.

Do not be mistaken. Larche is not St. Francis of Assisi. No one finds a gold mine in the cagey, deceitful, bullshit-flying world of prospecting without being able to recognize that not all of God's creatures sing songs of unadulterated truth. But that is just part of the job, not part of John Larche.

Larche meets you at the door. He is a tall man, six feet, and sturdy. His face is smooth, not a wrinkle on it. His hair is thick and dark, greying only where it edges into thick sideburns. His eyes are green, set wide apart, and utterly still. Their regard is not unsettling, but if you had arrived with a vacuum cleaner for sale you would probably decide to forget it and try next door.

The living room is large, bright and clean. Morning sunshine pours in through the picture window and the gold broadloom glows warmly. The room is panelled, and the walls are hung with seascapes and landscapes. Larche reclines in one of those heavy chairs that tilts back when you push on the armrests. He is wearing a beige shirt, brown trousers, brown socks and sandals. This is not a man whose head has been turned by riches.

The prospector's job is not the discovery of gold mines. All prospectors would like to do just that, of course, and if they do it is always fair sport to take full credit for it, as if some unerring nose, some gift, some special sense has led them to that place in the bush where the gold throbs its song. But really, the job of a prospector is to head into the bush and look for prospects. A prospect is simply a likely looking piece of ground. You scrape away the overburden, or ground cover, to get at the rocks. You use your backhoe to do this. Then you bang a few holes into the rock, blast, pick up the pieces, take them out and try to flog them to someone in Toronto with enough money, savvy and greed to be in the mining business.

"It's really not that hard," says Larche. "It's not that hard to get some rocks together and have an engineer write a report that says there might be something there. After all, there might be. Who can say for sure?"

But he smiles as he says this. For of course it is not that simple. If it were, the woods would be swarming with awful little people from Toronto clad in Eddy Bauer bush gear and cracking their shiny new backhoes into one another as if it were a sleet storm on Highway 401. No, you have to know what to look for in there. You cannot slap any old piece of granite down on a geologist's desk and expect him to write a report saying it looks promising. He might say it looks promising if you are going to open a granite mine. But he will not say it looks like gold unless he finds some of the very stuff leaking out of what you bring. And even then there is no guarantee that people will listen very hard. Take Hemlo, for instance.

"I knew about the area for quite some years before I actually went down there. Everyone in the mining business around here knew about that area. You could hear the talk. There's always lots of that, of course." Larche smiles. It is like saying there will always be flies.

"You talked to people who had been there, maybe who had prospected the area. You looked at the records of the work that had already been done. All that is on record. All you have to do is look it up. There were a lot of people who'd worked that ground, and all of them . . . they gave it a good shot." Larche pauses as he adjusts his chair a fraction. Then he looks off into space, a little dreamily, and nods.

"But they missed it."

Yes, they missed it, and so they die unknown. Or they die broke. Or they die still searching. Or they die wondering where it was, and wondering what if. What if they had tried over that last hill, just where the fault began to penetrate that different-coloured rock. What if?

"This was work — what had been done down there — this was work done by competent guys. And you know . . . they came up with gold down there before Don McKinnon and I ever got near the place.

They came up with gold years ago. But it was erratic. It was just too erratic. Nobody would go for the property with showings like they found. Nobody. Too bad.'' Pause. ''Well, for them.''

The whole business of a gold strike like Hemlo is a web of fluke. Back in the 1950s, when some of the work that Larche refers to was being done on the Hemlo properties — and being done by people who knew the difference between gold and pyrites — gold was not a very glamorous metal. It was trading down around $35 an ounce, and the price hadn't shifted since the 1930s. When men went into the bush, most of them were not looking for gold in those days; they were looking for copper and nickel. Back then the prices for copper and nickel were climbing. Gold was the wallflower at the dance; she just sat there.

Men had been exploring Hemlo as far back as 1869. South and west of the current activity, near the present town of Heron Bay, an Indian prospector by the name of Moses Pe-Kong-Gay discovered two veins with substantial gold. He showed his discovery to two developers, who interested a mining company to the extent that a number of shafts were actually sunk and some gold ore was shipped. Lytton Minerals Limited holds the option on that property today.

In the 1920s the station agent at Hemlo, Jacques LeCour, an inveterate prospector, sank a number of test pits on property that is now optioned to Bel-Air Resources. That property is just north of the old site of the Hemlo station, right on the Canadian Pacific Railway main line. At the same time, a number of trenches were being dug to the north of the line, on property that now belongs to Golden Sceptre.

Men were trenching right through the heart of the Hemlo goldfield sixty-five years ago.

In 1931 the whole region was mapped by J. E. Thomson of the Ontario Department of Mines. He recommended two areas for exploration: the area that surrounds the present town of Manitouwadge, well to the north of Hemlo, and the area that is being pierced today by three major shafts. Later, Peter Moses, an Indian prospector from the reserve at Heron Bay, made a discovery in the western sector of what is today prime Hemlo ground. He showed the property to a Mr. Ollmann of Heron Bay, and the property was staked in September of 1945 by Dr. J. K. Williams of Maryland, apparently in partnership with Ollmann. Williams staked a total of eleven claims. This became the ground known today as the Williams Option, a block of patented claims. Patented claims do not have to be worked on a continuing basis in order for the rights to the minerals to remain with the owner. A patent in effect grants the holder outright title to the land, as long as he pays the taxes.

The adjacent ground, now optioned by International Corona Resources Limited, was staked by Moses and Williams in association with three other prospectors, all of whom formed the Lake Superior Mining Corporation Limited in 1947. The company proceeded to develop its ground at Hemlo for four years, mapping, trenching and diamond drilling. And they were so close. They were probing around the very edge of the Hemlo ore body. But they missed it. And they did what ninety-nine men in a hundred would do. They gave up.

In 1951 the property was optioned by Lake Superior to Teck-Hughes Gold Mines Limited. They began to strike gold. Teck-Hughes drilled six thousand feet and defined *89,000 tons grading 0.27 ounces of gold per ton*. And then they walked away from it.

The question is: Is it ore? To a mining company, minerals are just so much rock until there is a deposit rich enough to make it worth mining. If gold is selling for, say, $35 an ounce, and it is going to cost $75 an ounce to take it out of a particular piece of ground and make it into gold, then that gold-bearing rock is not ore. To be ore, it must be capable of being mined at a profit. That is the definition of ore in the mining business, and that is why Teck-Hughes walked away from Hemlo after drilling the property again in 1957 and 1959.

The Teck-Hughes partnership dissolved in 1965, though both members are back on The Hemlo now. Teck is the major tied into the Hemlo gold play with International Corona. A major is a mining company with the fiscal muscle to explore a property thoroughly. A junior company is usually a promoter. And Dick Hughes is back on the gold camp, too, this time with his partner, Frank Lang. Like all of the north, the Hemlo has been watched through the years by the same wily faces, the veterans, the searchers. Only the paper entities change.

The present Corona property was next staked by four different prospectors in turn, but no work was performed. Presumably the prospectors tried to option the claims and failed. A prospector options a claim when he sells an interest to someone who will undertake the exploration.

Time ticked another few years off the clock, and the gold sat right there beside the Trans-Canada Highway. You could have stopped your car, walked over to a rock cut, chipped a few pounds off and stuck it in the trunk. In 1973 a company called Ardel Explorations Limited staked the property, sank three holes and revised the tonnage. It was the same zone still. There were, they now judged, 150,000 tons grading 0.21 ounces of gold per ton. Not enough. This is why.

A deposit of 150,000 tons grading 0.21 ounces of gold per ton yields 31,500 ounces of gold. Say the gold price is $400 an ounce. Then the deposit is worth $12.6 million. This sounds like a substantial sum, and it is. But to get that gold *out* would cost, even with an inexpensive open-pit mine, at *least* $20 million. Conclusion: the gold in that deposit is not ore.

The company cancelled its charter. The ground was restaked in 1976. But in December of 1979 the claims came open.

The Hemlo ground was up for grabs.

"In spite of all that work, the area was just not known as good gold country," Larche says. "I think part of the problem was the drills they were using then. They used a lot of X-rays, and you don't get as good an assay from them; it's not as representative as the drill cores today."

The standard width of a diamond drill core today is an inch and a quarter. The X-ray drills Larche refers to were only three-quarters of an inch wide. Water was pumped into the hole to wash the cut rock away from the spinning face of the drill and force it back up the hole and out of the way. According to Larche, the water may have washed enough of the gold away that it affected the assay, grading the rock lower than it really was.

That might have been part of the problem, as Larche says. But it wasn't the whole problem. The whole problem was that they missed the best gold. Period. They looked, and they didn't find it. Later someone else looked and did find it. That's the way it works. That and one other factor. In places like London and Zurich and Paris and Hong Kong, people were starting to pay a lot more for the metal. That makes you look again, and this time you look closer.

"You know, a prospector doesn't have to take a property all the way to patent; he doesn't have to do enough work on it to keep it for good. Once you have a prospect that's sexy enough, the average prospector takes it to a major or a junior mining company and he makes a deal. What you want is, you want someone to *explore* the property, you want them to dig. In most cases you're looking at hundreds of thousands of dollars to do that exploration. In the case of Hemlo, we already had reports of previous work, there had already been a lot of drilling. The work indicated a very attractive environment for gold."

A little smile plays across Larche's face. "It didn't hurt when the price of gold took off. Around the time I started to take a look at Hemlo, gold was breaking past $500 an ounce. This changed the appeal of the area."

The surge in the value of gold is the result of a curious sequence, but a simple one.

A group of businessmen in Zurich and London may watch the world with alarm. It is in the very nature of these businessmen to distrust currencies. That is their job. Their loyalties are to ledgers, not nations, and they move around their resources, or other people's resources, accordingly. If they do not like what is happening to paper money, and they want a refuge for their wealth, and they figure that refuge is a certain commodity with "999.9 fine" stamped on the side, then that is what they buy — gold. When that happens, a lot of rock in Ontario will become ore.

You see that coming, you become a rich man.

"We knew we had what you'd call a very good prospect. At least, I did, and I guess Don was thinking the same thing at the same time. The price of gold went up fast. There had been an engineer's report, and we found it. It covered the property that Corona is on now, and part of the Golden Sceptre ground. You don't have to actually go to the ground to get those reports. They're on file. This one was at the district mining office in Thunder Bay. All the research I had — that's where I got it from.

"Then I realized the ground was open."

What Larche realized, and what McKinnon had stumbled across at the same time, was that the claims on the Hemlo property had lapsed. There had not been enough work done on the property, and the claims had been cancelled by the district recorder. All you had to do was walk into the bush, stake those claims with wooden stakes, ink your prospector's licence number onto the wood, and walk out again with the right to sell some mighty interesting ground. Then Larche started to hear rumours.

"Well, I heard that Don had already been down there and staked all the ground. If that was true, then there'd be no point in me going down. There'd be nothing there for me to stake."

Larche looks over, and a glint of amusement is there.

"That's what I heard, but with Don you're never sure."

What Larche alludes to is the possibility that the rumour that Don had already staked the Hemlo property may have originated with Don himself. It could be a bluff. In fact it *was* a bluff, if it originated with Don. For he had not staked all the Hemlo property. Larche at that time didn't know for sure. But he did suspect.

There is no rancour when Larche relates this tale. It is simply a *ruse de guerre* that he is talking about. As he himself readily admits, the rumour that you have already staked a property is just a technique that a prospector uses to keep other prospectors away until he *does* have time to stake the ground in question. If the ground is a thousand miles away in the bush, maybe the other guy will think twice before he fires up the plane and heads out. Maybe he'll believe you, maybe not.

"My son David and I, we just piled into the truck and drove straight down there, three hundred miles, no stopping."

It was 27 December 1979.

Two

The Partnership

John Larche was fifteen years old when he took his first mining job. You were supposed to be eighteen, but the year was 1943, and with a lot of the miners away at war, no one looked too closely at the papers presented by the strapping lad who walked quietly into the hiring office at the Preston East Dome mine. Larche worked underground with the diamond drill team. They would work out along the horizontal tunnels — called crosscuts or drifts in mining jargon — probing out into the rock to see whether the vein currently being mined might be paralleled by another vein. Larche worked underground for five years, and then abandoned the subterranean world for surface exploration.

Work underground is a damp and grimy business, and there is always the sobering rattle of falling rock. This is a constant in mining: little chips, and sometimes large pieces, are always falling from the roof of the drift. At strategic areas, where there is valuable electrical equipment, wire mesh is bolted to the roof to keep the loose rock off the gear. Elsewhere, rock bolts are driven into the stone to secure holding plates. All of this is supposed to make the drift a safer place. What it does is remind you of how many tons of rock are piled above your head. Of course, work on the surface has its own hazards.

''One time I was out in the bush doing electrical work. You carry this thing like a hula hoop. I had no ax with me, and no gun. I should have seen this big mother bear coming, but I had my head down. So did she. And the wind was coming from her to me, so she didn't smell me, either. We were practically eye-to-eye when she finally saw me. She let out a whoop and her cub took off up a tree. She stood straight up. There was about twelve feet between us. She swiped at a poplar and tore it right in half. She was frothing at the mouth and advancing. You don't have to know the language to know she didn't want me around.

"I backed up very slowly. I didn't want to aggravate her or frighten her in any way. By the time she stopped advancing there was no more than six feet between us. She could have just reached out and cleaned my head right off my shoulders. But she went away."

Nowadays Larche rides into the bush on a $100,000 muskeg tractor-and-backhoe rig. With this he scrapes off a few yards of overburden to get down to the bedrock, drills, dynamites, and takes a closer look. If he likes what he sees, he might stay and dynamite some more. If he doesn't, he's off to another claim.

Don McKinnon is the scholar on the team. He's a year younger than Larche, shorter, and squints through a pair of glasses with a look of permanent curiosity. When Don McKinnon looks you over, you may be forgiven for feeling a little like a piece of rock in which McKinnon doubts, but nonetheless hopes, that something valuable may be found. His research library is legendary in the northern mining community. He keeps track of areas that have been explored and abandoned, and finds no reading so engrossing as a stack of old assay reports and geological surveys. So profuse is this mass of paper that it has overflowed his farmhouse in the tiny community of Connaught and is now threatening to engulf his townhouse in nearby Timmins.

McKinnon comes from a family of railroaders from Cochrane, Ontario. There was nothing in his early years to mark him for success. "In fact, I quit school well before the end of high school," McKinnon recalls, adding with a chuckle, "The principal told me I'd end up digging ditches, and he was dead right. It's just that I find gold in them."

Soon after leaving school, McKinnon moved to Timmins and started to work in the bush as a supervisor for a logging company. Since the job took him into the bush anyway, McKinnon decided he might as well learn something about prospecting, and soon began to stake claims of his own on the weekends. By 1963 he'd had enough of logging and determined to become a prospector full time.

"I wanted to be a millionaire by the time I was forty," McKinnon says. "And there's a damn good chance I'd have made it if I hadn't concentrated so hard on blowing it. Whatever I made I spent. Convertibles. Stupid things." But don't get the idea that there's anything about those years McKinnon regrets: "In the bush you work hard, eat lots, put your tent by a nice lake, have a dip, and by 10 P.M. you're in bed."

Larche and McKinnon were not partners when they began, independently, to make their secretive appraisals of the Hemlo property and head into the bush for the final swoop. Each man in his own way had followed the threads of his experience and his instinct, and those threads had taken him to Hemlo. Each man had listened to the tales of old prospectors, men who were certain there was real gold at Hemlo if a man could just find the right tip of rock, the right hollow, the right hint glittering in the clear waters of a lake.

When Larche and his son David arrived at the Sharl Inn Motel at Manitouwadge Corner on the Trans-Canada Highway, McKinnon was already there. But he was off staking in the bush with his two sons, Duncan and Don.

Larche smiles. "I guess he didn't have the whole property staked, or he wouldn't have been out in the bush. It was pretty cold.

"My original intention when I headed down there was to stake eighteen claims. But when Don came out of the bush, I found that he had already staked twelve claims on what was then thought to be the favourable geology. That was right where I had wanted to stake myself.

"But I told Don that I wouldn't stake around him. I said that instead, I was going to go over and stake the old Lake Superior property, a little to the east. Don said I couldn't, that the property wasn't open because he'd already checked. Well I said it *must* be open, since those claims were staked and worked at the same time as his claims had been staked and worked. Both properties would have been forfeited at the same time.

"We disagreed.

"The next morning David and I went in to stake the Lake Superior claims. Meanwhile Don called the mining recorder in Thunder Bay, and he found out that I was right."

There is a lot of activity these days where the road from Manitouwadge meets the Trans-Canada Highway, but in December of 1979 it was the definition of bleak. Twenty-five miles down the road one way was the paper mill town of Marathon. Twenty-five miles the other way, and you come to White River, a place which actually *brags* about having the coldest recorded temperature in Canada, presumably below the Arctic Circle.

In December of 1979 all there was at Manitouwadge Corner was a gas station and the uninviting structure of the Sharl Inn Motel. There was nothing to suggest that two strong and wilful men were scrambling through the freezing, cracking bush with their sons, slashing at trees with their axes and scribbling their licence numbers across the cuts as fast as they could. Out on the highway, the big semis swept past on their way to the Lakehead or the Soo, but deep in the bush the axes rose and the axes fell as two men blazed the boundaries of the richest, fattest, most thrilling goldfield the Americas had ever seen.

The men didn't know it themselves.

"All you can really say," Larche insists, "is that we knew the property deserved a damn good look. I've been in this business forty years — thirty years of it prospecting full-time in the bush — and I can say this property looked damn good. But I would never have said it was what it turned out to be."

And it was good enough that Don McKinnon was taking no chances at all, but was on that phone to the mining recorder as soon as he could get through to find out for sure whether Larche was staking ground he could keep.

And he was.

"It turned out that this group I staked was pretty sexy. The government was reporting at least eighty thousand tons on it grading 0.27 ounces of gold a ton. I staked seven claims there the first day, and those claims were every bit as impressive, if not more impressive, than Don's claims. Well, Don and I talked about it later at the motel. This was after he found out I was right, eh? As it turned out, he liked my claims and I liked his claims. So we pooled them. We just struck a deal right there. Fifty-fifty."

This is the manner of deals in the bush. Two men meet one another coming over a rise, and they're staking the same terrain. They can do one of two things. They can go to court and waste their time and make some lawyers rich, or they can shake hands, keep staking and spread the risk. That's how Don McKinnon and John Larche agreed to divide the gold play. A handshake, and keep staking.

There are a lot of little twists and turns in the gold business. Some of them make a man poor where he might have been rich. One such twist was letting the Hemlo claims lapse. Whoever did this has probably been butting his head into a wall ever since the gold play broke in earnest.

Another twist was the superstition of John Larche about the number thirteen. This was a good twist.

"It was around March of 1980. Don and I had decided to do a little more staking at Hemlo. We went in and we staked six more claims around the original seven that I staked with Dave in December. Later, when I was sketching in our staking program on our map — sketching in what we'd staked — I noticed our claims added up to thirteen on that part of the property. I mentioned it to Don, and we just headed right back in there and staked another.''

That fourteenth claim is smack in the middle of The Golden Giant, the site of some of the richest ore at Hemlo. If you had gone up to The Hemlo yourself with the thought of staking only one claim, that fourteenth claim would have been about as good, solid and rich a claim as you could have picked. Your interest in that one claim would put about $500 a day into your pocket, 365 days a year, for about twenty years. Almost $4 million. Not bad for a half-hour's work.

Of course, it is important to remember that Larche and McKinnon were not simply romping through the woods, pounding their stakes into ground that jumped and glittered with the fiery signal of freely occurring gold. They were staking land that had been staked and staked and staked again. They were scrawling their licence numbers around property that others had written off. But interest was growing now.

The Geology of Gold in Ontario is a scholarly work published in 1983 as Miscellaneous Paper No. 110 of the Ontario Geological Survey. It is almost three hundred pages long, complex and detailed. Most of it is incomprehensible to the layman. But there is a relatively short section by the editor, A. C. Colvine, where the unversed need stumble only once or twice. It is worth quoting because of the plaintive nature of the conclusion, where for a moment the scientist drops his mask and reveals the puzzled, human face.

> The recent discoveries in the Hemlo area have aroused considerable interest and new hope that there are many substantial new deposits still to be found even in areas which have a long history of prospecting and exploration. A striking characteristic of the Hemlo discovery is that it is so readily accessible; an outcrop of the ore zone was exposed in a road cut during construction of the Trans-Canada Highway more than twenty years ago; there are obviously many less well explored or accessible ''greenstone'' sequences within the Shield.
>
> Although exploration for gold in the area dates back more than one hundred years and significant discoveries were made, the much higher potential of the area was not recognized until recently, probably because gold occurs in an innocuous looking unit. . . .
>
> [Ontario geologist T. L.] Muir mapped the Hemlo–Heron Bay belt in

> 1978 and 1979. In his initial report he presented assay data from *a previously unreported occurrence and recommended the Hemlo area for further gold exploration*. Muir describes the general geology of the belt based on his previous work. In addition he describes from his work and available published information the geology around the Hemlo deposits. Most importantly he points out that this is a geologically complex area and needs a more thorough field study before all of the geological features can be determined.
>
> The Hemlo area has highlighted the fact that visible gold, particularly in quartz veins, has been the principal prospecting target. While there is visible gold at Hemlo, examination of some surface material from the deposit leaves the question in the mind of the observer, "How often have I seen this type of rock and not even thought to sample it for assay?"

The italics are mine. They are there to show that other trackers were in the forest with Larche and McKinnon, and they were stalking the same quarry. But Colvine's human question at the end still rings with all the eternal puzzlement of man: Why did I not see this when it was there before my searching eyes?

"We knew there was *some*thing there."

John Larche is recalling what he and his partner were deciding after they staked their first twenty-six claims. "Who do you go to with your ground? Who should you try to option the ground to? A junior mining company is usually better to deal with for a prospector. They give you more cash and they give you a better deal. A major has to ignore its own bureaucracy to deal with you. Somebody has to decide that all the money they've spent on their own explorations, all the time and study they've put into their own looking, their own staking, their own options — somebody has to decide to ignore that and deal with this guy who comes in from the bush with dirt under his fingernails."

Larche just shrugs his shoulders.

"A junior now. A junior has no prospecting crew or staff. They are reliant 100 per cent on prospects being submitted to them."

Larche was determined that he and McKinnon would find someone who would give the ground a careful look. He was determined to find someone who would actually bring the diamond drills onto the property and start to pull cores.

"Not that we thought we'd found the richest ore body in the world. Not by a long shot. But we thought we had something as good as maybe Detour Lake. We thought we had a gold mine there. And so we wanted a company that would do a lot of drilling. We thought the prospect looked extremely good, and that the amount of gold indicated looked good. We felt that what was maybe not feasible twenty years ago was feasible today.

"And we absolutely wanted a company that would guarantee an extensive program of diamond drilling. You know, Dave Bell agreed with us 100 per cent that there was a good ore body there."

Dave Bell is a geologist, and the man who really found the Hemlo ore. When Vancouver investors finally decided to explore the site, it was Dave Bell who directed the diamond drills that at last started to pull those dazzling cores of gold-laden rock out of the Hemlo ground. Before that, he had helped Larche and McKinnon go after financing. Bell's office is on the second floor of a small office block behind Timmins's Empire Hotel. It is a fitting place for a geologist's office. The Empire is the repository of more tales of gold rush and copper rush and fast, fast millionaires than any hostelry in the north.

It was in the old Empire that John Larche and the mayor of Timmins, Leo Del Villano, pulled a stunt that got their faces in every paper in the country. The year was 1964, and Timmins was busting at the seams with millionaires. The city was fat and wild and flush as a drunk on payday, for 1964 was the year of the Texasgulf Inc. copper strike, a mineral discovery that turned a couple of hundred square miles of moose pasture into the hottest ground in North America. The magnitude of the copper strike was sending men into the bush with nothing more than an Esso road map and a pair of new boots. The woods were teeming with greenhorns, and the streets of Timmins had never been littered with a sadder collection of fly-savaged refugees.

Hounded by predatory tipsters, sucked dry by merchants and mosquitoes both, miserable and greedy by turn, these desperate men stumbled around in the bush for months, knocking over staking posts and generally making themselves a nuisance to the people who were in there to make money. Like John Larche and Don McKinnon.

"We made some money, I guess," Larche remembers. "And one night we were all sitting around in the Empire, partying. I think Murray Pezim was there, too, you know. He was in on the Texasgulf strike. Hell, everybody was in on the Texasgulf strike. We were sitting around and talking to these reporters, when the mayor offered to light a cigar with a $100 bill. They liked the idea, and the photographer got ready to shoot a picture.

"But I told him I thought it was illegal to just light money on fire like that. So instead we sprayed the bill with water so it wouldn't burn, held a piece of toilet paper behind it, and lit the toilet paper. It looked great. It looked just like the bill was burning. A Canadian Press photographer took the picture and it moved right across the country.''

Proximity to the infamous Empire notwithstanding, it is difficult to imagine Dave Bell getting up to stunts with $100 bills. Bell is a studious-looking man in his early forties. He is a little owlish behind spectacles, but fit. There is nothing cramped about these offices; the secretary could probably stay in shape just from the trip back and forth to the filing cabinets. Today she is hunched over a dictionary of geological terms. Under the desk, her shoes are off. There is a percolator burping out the aroma of fresh coffee, and light floods in through the big windows. This is the kind of office everyone would like to have: enough room to stretch your arms or have a good pace. In Toronto you have to be the chairman of the Royal Bank to get this kind of office.

There are maps on the walls, and a large, picturesque calendar. Piled into a corner near the stairs are a couple of very worn and sturdy packsacks.

''That's where I really like to be,'' says Bell. ''Out in the bush.''

The paraphernalia of office geology — maps and surveys and heavy, dusty books — clutters the top of a long credenza. Bell motions his visitor into a chair and slumps into a high-backed executive throne that looks incongruous in these easy precincts. He is remembering the prospect that John Larche and Don McKinnon brought to him.

''Sure, The Hemlo.'' Bell shakes his head. ''The Hemlo. Well you know, quite frankly, it didn't look all *that* hot at first.''

Bell tilts his chair back and stares at the ceiling for a minute, recalling.

''The main reason, I think, that Hemlo was never discovered, discovered for what it was, is that it's so *different*. Traditionally in this country we've looked for and found gold in quartz. You can get a quartz vein and there it is, chunks of gold in there that you can actually see. Probably what happened with those earlier prospectors — you know, that ground had been gone over and gone over a hundred times — probably they just didn't know what they had. Put another way, they really didn't know what they were looking at.

''Now, the only thing that John and Don brought me were those earlier reports, reports of geological back-up work. Those reports indicated a low tonnage of marginal ore. The only thing I saw that improved the prospect was that I knew there were people willing to look at these marginal properties.''

Bell thinks for a moment.

''But it wasn't going to be easy. A lot of those people in offices in Toronto, they'd been driving right past that property for years. It was right on the Trans-Canada Highway.''

What Bell is alluding to is the reluctance of the mining community to believe that a good mine can be located anywhere but a couple of hundred miles into the bush. Unless you have to build a really expensive road to it, well then, how can it possibly be a good site? This is an endearing characteristic of human nature only if you are not trying to flog a prospect on Bay Street. If you are, it becomes the kind of reaction that makes you want to grind your teeth and pound on a few thick heads.

This whole reticence on the street — the unwillingness to believe that a gold mine could ever be found simply sitting alongside the Trans-Canada Highway — is part of the reason that Toronto missed out on much of the early action on The Hemlo. Peter Brown, formerly a chairman of the Vancouver Stock Exchange and now the president of one of Vancouver's leading brokerages, puts it this way:

''Look, that mine was sitting right on Toronto's doorstep, and they wouldn't see it. Not at first, anyway. That's how it got taken out from under their noses. By us. Hell, talk about a good location! One of the major cost factors in developing any new mining property is always transportation.''

Brown is scornful of the big shooters in Toronto who missed the early Hemlo play. But he's gleeful, too.

''Transportation? You want to ship the ore out of Hemlo, you could put it on the bloody Greyhound!''

All of this is obvious now, but it wasn't obvious when John Larche and Don McKinnon went looking for someone to buy their prospect. Otherwise they would not have encountered such a uniform wall of indifference. Up and down Bay Street they plodded, knocking on old, familiar doors, and knocking on new doors, too. Sometimes the doors opened. A lot of the time the people behind the doors were ''out.'' On the street, it doesn't take very long for the word to spread that a couple of the boys are down from the bush trying to beat a dead horse into a couple of grand. On the street there are more people asking for money than there are people handing the stuff out.

Three

Selling Hemlo

The first strike at the Hemlo bait came from a thirty-five-year-old promoter named Don Moore. When he caught wind of the Hemlo staking early in 1980, Moore was working on a five-year contract in Kirkland Lake as regional industrial commissioner. His job was to attract development to the municipality, and that is what he did. Under Moore's stewardship the town gave itself a facelift and began to talk itself up in the right circles. Federal and provincial money trickled in, and hard on its heels came the prospectors, geologists and promoters who unlock the investment dollars that fuel the northern economy.

That is what Don Moore did for his daily bread back then in 1980, but you cannot buy a promoter with a salary. A promoter's ears are always straining at the wind for the rumours of a new prospect. A promoter is always dreaming, and in Don Moore's case the dream was a dream of gold.

Don Moore grew up in a gold camp. As a child in Red Lake, Ontario, he lived among some of the richest mines on the continent. He heard the miners' tales and listened to the prospectors talk. ''All I really want in life is to run a gold mine,'' Moore frankly admits. ''It would be nice to own it, too.''

Moore is sitting in the living room of his apartment in Toronto's swank Manulife Centre as he talks. He is darkly handsome, thick hair brushed back from his forehead. He is dressed in a sweater and slacks, his stockinged feet crossed on a heavy coffee table made from what appears to be a huge block of worked metal. From one corner, music streams from an elaborate sound system. The view is north. It commands nothing more inspiring than the grey tower of another part of the complex. Still, people pay enormous sums of money for the privilege of being able to stare at other buildings in this part of Toronto. The Manulife Centre is at the corner of Bay Street and Bloor, and if you had a middling throwing arm you would be able to put a rock through the windows of five or six of the most expensive stores in Canada, just from that corner. This is pricey turf, all right, but it is not where Don Moore's heart is.

"I don't even keep a car down here," he says, staring out the window. "I keep one up north. That's where I need it."

When Dave Bell told Larche and McKinnon that he knew of certain people who were willing to look at marginal properties like Hemlo, Don Moore was one of the people Bell was thinking of. The mining business in Canada is an infinitely complex web of contacts, and depends as much upon the friendships a man strikes up as it does upon the rumours and the claim-staking and the drill results it trades in.

"I first met Dave Bell in 1968 in Thunder Bay. I had just graduated from the University of Waterloo, and I had a biology degree in one hand and a piece of chalk in the other. I was marking the boards at the Midland-Doherty office, learning the brokerage business from the bottom up. That's where I met Dave. Thunder Bay's a small town, and all the guys in the mining game would drop in to check the market."

When Dave Bell called Moore to tell him about the prospect that Larche and McKinnon had for sale, Moore knew just the man to call. Steve Snelgrove had been a classmate of Moore's at the University of Waterloo, graduating with a degree in biology as well. But neither man was cut out to be a biologist and Snelgrove, too, entered the business world. It was to Snelgrove that Don Moore turned in 1980 when he heard about the Hemlo claims.

"Steve had a company he was promoting at the time. His plan was to option the claims and steer them into that company."

Moore called Don McKinnon at his home in Connaught and made an appointment to drive over and see him. Connaught is ninety miles from Kirkland Lake, but with gold in the air, Moore would have driven twice that distance. The two men met at McKinnon's home and sat down to talk business and rough out an initial agreement.

"I think it was just sketched out on a piece of paper," Moore recalls. "The meeting was not to establish a formal agreement but to set general parameters which Steve would find acceptable. We were just sitting and working it out at Don's kitchen table. He was there with his son. Don's place was quite modest then. When we were through, I just drove straight back to Kirkland Lake, in the middle of a snowstorm, and called Steve.

"Part of my promotion to Steve was that I would deliver a really good geologist. That was Dave Bell."

The wheels turn within the wheels. Dave Bell learns about the Hemlo claims and mentions them to Don Moore, whom he knows is watching out for a new property. Moore leaps, and nails down a preliminary agreement for Steve Snelgrove. Moore and Bell had known one another since 1968, a relationship of twelve years. Moore had known Snelgrove since 1965, friends for fifteen years. Bell promotes to Moore. Moore promotes to Snelgrove. Snelgrove promotes in Toronto. Moore agrees to deliver a good geologist — Bell. Here is the Canadian mining industry at work in microcosm. Friends talk; deals are born. The network clicks and shuttles.

"These types of deals happen every day of the week somewhere in Canada," Moore says. He is talking about the apparent haphazardness, the casualness, the handshake-and-a-nod atmosphere that seems to dominate the first stage of deal-making in the mining business. "A lot of these things never get tied down till they get to the lawyers. Everybody wants to keep it simple — a simple deal. Because once it gets to the lawyers, your costs are going straight up.

"Deals are ever-evolving; a deal evolves according to the circumstances; you are always making amendments as you need to. Hemlo was no different. Don't forget, nobody knew then that we were looking at a $5-billion or a $10-billion gold mine. It was just a prospect, and I had someone looking for a prospect. The nice thing was there had been a little work done, and so there were some figures we could put to it."

Steve Snelgrove is a tall, handsome, ruggedly built man, a college athlete, made for self-confidence, smartly dressed. If you were to see Steve Snelgrove march past on the street, the smell that would come off him would be the smell of success. That was four years ago. You mention Steve's name on the street now, you might as well be talking about Mother Theresa. There is nobody down there at the corner of Bay and King streets who will admit to even knowing who Snelgrove is. Or, if they know, *where* he is.

"I don't know. Steve Snelgrove? I don't know. He could be in the

States. I *heard* he was in the States,'' financier Claude Bonhomme mutters distractedly when asked about the former wheeler-dealer.

Snelgrove was a fast-lane moneyman and general deal-spinner on the street when Larche and McKinnon were down trying to peddle the first block of Hemlo claims. The two prospectors were getting mighty footsore when they finally hooked up with Snelgrove, and when Snelgrove began to talk, it sounded like music.

He was, the prospectors say Snelgrove told them, in the midst of a takeover campaign that would net him and his associates the control of a large company. This is not as fantastic as it sounds. There are people in Toronto who are quite adept at taking over other people's companies, and sometimes they even manage to do it with the target company's money. Such wizardry is possible. But not everyone manages to pull off these takeovers. If they did, then the corner of Bay and King would be nothing but a mob of unhappy executives milling around trying to find out who was in charge.

When his takeover was complete, according to Snelgrove, then the development money for the Hemlo claims would be available. He would fund an exploration program on the claims himself. This is how the story went. Alas, it went no further than the good-yarn stage. It became apparent to the prospectors that Snelgrove's attempt to gain control of the target company would fail. And since it was from the target's treasury that the development money would come, the skies looked a little dark for Larche and McKinnon.

Snelgrove proposed another tactic: hit Vancouver.

The Vancouver Stock Exchange runs with cash. It is the great speculative exchange of the country, the place where the risk-takers congregate to try to sell one another their fancy pieces of paper. Occasionally, too, somebody slaps down real money and goes after a property with drills and a heartful of hope. If there were no exchange in Vancouver, there would be no mining business in Canada. That is the belief of many people in the business, and even the most conservative mining czar holed up in his bank tower in Toronto will admit that few marginal properties would ever make it to exploration without the gutsy crap shoot of Vancouver to feed them the cash. And all you have to do to turn a marginal property into a hot property is pull a few drill cores out of the rock with the right assay.

Why are the speculators, the bust-'em-loose slapdash traders of the penny stocks, holed up in Vancouver?

Until the mid-1960s, junior mining stocks, even the infamous penny stocks, were listed on the Toronto Stock Exchange, and you could trade Fly-by-Night Mines on the big board right along with such sturdy matriarchs as Noranda, the grand old lady of the rock chewers. But a scandal struck the mining community, and somebody had to pay for it.

When in 1964 Texasgulf Inc. struck a rich copper field outside Timmins, the town went wild. If you take the clattery old bus that runs from downtown Timmins out to the mine site at Kidd Creek, you will get some idea why. Two great headframes tower a thousand feet into the northern sky. Below, trucks groaning under seventy tons of rock apiece beetle about at their ceaseless toil. For miles in every direction, the blasted and bare-scraped desolation stretches. And in the centre of it all there is the great pit. Rockslides are filling it in now, but no matter. The Kidd Creek mine moved underground long ago, and the empty, booming space of the old open-pit digs is simply left as a reminder of the tiny scratchings that took off the first ore.

There is a model of the Kidd Creek mine in a display room at the site, and it helps the visitor to understand why the strike was so explosive, the lode so rich.

Sitting on a table is a model of the downtown core of Toronto. All the big bank towers are there, and the CN Tower, too, a tall needle pointing straight up. Everything is built to scale. In the same scale, but suspended above the puny city, is the ore deposit of Kidd Creek. Put it like this: if they were to drop that copper deposit onto the downtown core of Toronto, the city would be buried far, far beneath the surface. In fact, even the tip of the CN Tower would still be a few thousand feet from the surface of the ore pile. When Texasgulf struck the ore in 1964, they knew it was very big, but they didn't know it was *that* big.

Anyone who could batter a length of wood into the ground grabbed a prospector's licence and headed into the bush to stake. They were selling claims so far from Kidd Creek that you'd have to own a plane to get there in a day. But that is the way of a staking rush. The aura that casts its enchantment over the strike spreads far afield, and the speculators and promoters trade the paper around so fast it looks like a snowstorm. One of the mines that got traded in this way was Windfall Mines. You mention Windfall Mines today in the dining room of any of Toronto's better clubs, you wait a long time for a table.

Windfall traded from a few pennies a share to $5.65 almost overnight. This made a lot of people happy. Not for long. When the assay reports came in, well, Windfall's prospects looked about as cheering as an empty street in the cold, hard light of dawn.

Windfall Mines went down like a dropped stone, and when it hit bottom, a lot of very angry people got off. Viola McMillan — before the fiasco one of the most respected members of the Canadian mining community and an owner of Windfall — went to jail. Suddenly the penny dreadfuls were unwelcome in Toronto. The Ontario Securities Commission tightened up the rules with a will, and all the penny stocks and the promoters who dealt in them went looking for another home. They found that home in Vancouver, where they were welcomed aboard with a slap on the back, and where they thrive today.

When his flight from Toronto came in over the delta and touched down at Vancouver airport, Steve Snelgrove knew just the person he wanted to see. Snelgrove's cab took the road that swept in from the airport and across False Creek, a narrow inlet off English Bay. Inland, False Creek is lined with the dazzling, futuristic shapes of Expo 86, the international trade fair. Elsewhere the shoreline groans under the weight of the furiously chic townhouses of the young professionals who can afford the extortionate rents. Across False Creek, the gleaming city rises against the dark green of Grouse Mountain, looming above it on the other side of Burrard Inlet. It is a glorious spectacle, Vancouver, and for an eager young tycoon storming in from the east, it is a place alive with two invigorating scents.

One of these is the tang of the good ocean air. The other is the smell of money.

At the corner of Dunsmuir and Granville, Steve Snelgrove got out of his taxi. The building he was headed for, 609 Granville, is right at the corner: the Vancouver Stock Exchange.

The exchange is housed in a tall, shining tower that drips with money and hype. Nailed down like a golden peg into the centre of the Vancouver financial district, it is the anchor that holds a mad gang of traders, the greatest assemblage of tub-thumpers in the country. This is tout terrain; this is where the bucket shop sits behind neatly lettered doors and cooks like a kettle on the boil. A bucket shop is an establishment where a lot of energetic people congregate to buy pieces of paper for fifteen cents and convince *you* to buy that same piece of paper for $15. They do this by trying to convince you that your $15 will turn into $30. And if you are greedy enough or stupid enough, you go home that day with a piece of paper worth fifteen cents. The tout goes home with $15. Cash.

Of course, there is always the chance that you will go home with something like Hemlo in your pocket, and you will be a zillionaire. That is why people congregate around exuberant, heads-up places like the Vancouver Stock Exchange. But it is wise to know who is who and what is what, unless you find your personal wealth too heavy to lug around by yourself.

The exchange is on the third floor. You take the elevator up, flash an Ontario driver's licence at the inattentive guard, who will assume it is a pass, and just walk in. The room is decorated in muted tones of green and grey. There is carpet on the floor. The traders are, for the most part, smartly dressed. A gallery runs around the room at mezzanine level, and along this gallery, back and forth, run the breathless boys who chalk the bids onto the blackboard that carries the trading symbols of all the companies listed on the exchange. Compared to Toronto, the chalk-and-run system is strictly Stone Age. But this is Vancouver, and what they do in Toronto registers about point-zero-zero-one on the Richter scale of Vancouver concern.

More than any other room in Canada, this is a room that breathes gold. As any trader bellowing his bid across the floor will tell you, gold is what makes the Vancouver exchange run. Say there is a tremor passing across the bullion market in Zurich, and gold dips a dollar or two an ounce. There is not a trader on that floor who will not know about it within thirty seconds of the change flashing onto a small screen in the corner. It is a fascinating place, but if you are inclined to enjoy standing around in wonder at the manifold works of man, then there are safer places to do it in. Beware of charging traders. This is the Vancouver Stock Exchange, not Westminster Abbey.

But the third floor is not where Steve Snelgrove was headed that day in 1980. Snelgrove was on his way to the fifteenth floor, where a pair of discreet doors announce to the world that these are the offices of Allcorp Management. To most people, this corporate name means nothing, and that is the idea. But to those who know how to translate such meaningless titular conveniences, Allcorp Management means Murray Pezim, and Murray Pezim is the monarch of mining promoters. Go ahead. Ask him.

When Snelgrove arrived from Toronto, Pezim was busy trying to juggle some unwieldy oil and gas deals he had at the time, and was unable to turn his full attention to Snelgrove. But what he saw of the property he liked, and he looked around for a loose company to park the deal in. He found instead Nell Dragovan. Murray had the bucks, and Nell had the company.

Nell Dragovan is the shadow lady of the story. She is a woman who operates with great deftness and self-confidence, but with such little holler and hype that it is difficult to believe she is in the same business as the rest of the promoters who crowd the Vancouver office towers around the exchange. Tracking Nell Dragovan is an exhausting and usually unrewarding exercise. She appears and vanishes from the scene with equal ease and silence.

Today another visitor from Toronto is standing outside the interior doors that bar the way into Murray Pezim's inner sanctum, just as Steve Snelgrove had done a few years ago. It is a pleasant, open space, secretaries clattering away on either side and the coffee machine spitting its little signals of gentle addiction into the air. The flow of traffic is ceaseless here. People troop into the inner office, some with hopeful expressions and others with expressions of dread; people troop out, some looking relieved, others crushed. Decisions are made behind those doors and they are made with a certain chilling despatch. But in the outer office, where the visitor is being shown his way to the private phone room, all is the brisk and cheerful robotism of the modern office. And the visitor asks: ''By the way, do you have any idea how I can get hold of Nell Dragovan? She's not listed.''

''Nell? Why, she's right there.''

The speaker turns and points, and there sits the object of such ardent pursuit, like a rare bird sought over half a continent to be found, finally, nesting in the back hedge. There is a long row of offices along one outside wall. The offices are separated from the corridor, and from this open space where the secretaries sit, by walls of glass, hung with sheer curtains through which the indistinct figures within can be dimly perceived. That is the first glimpse of Nell Dragovan: the indistinct shape of a woman sitting behind a desk, gazing out through the veil that conceals her. She has heard her name spoken, and she is looking to see what it is all about. Not with much curiosity really, rather a kind of still appraisal. She is watchful, but not eager. There is plenty of time, and doubtless she will find out why her name has been mentioned. She does not wait for long, but turns back to work, penning notations in the margins of a sheaf of papers positioned neatly before her.

Nell Dragovan was a legal secretary for a Vancouver law firm when it suddenly occurred to her that there was nothing particularly remarkable about the procession of young hustlers traipsing in and out, setting up companies, cutting deals and making bundles.

''It just seemed to me, finally, that there was no real difference between those people and me, except that they knew how it was all done. They were no smarter than I was; they had no access to magic information. They just happened to be in the market and doing what they were doing. At some point, someone had shown *them* how it was done."

Nell Dragovan determined that she would herself learn how it was done.

There is softness in Nell Dragovan's face, but there is an iron resolve there, too. Pale eyes, pale skin, pale hair, and a control so total that it can be a little unnerving. That is Nell Dragovan, and sitting in her office under that unblinking regard, trying to prise a few answers out from between long and utterly self-possessed silences, is not easy work. She sits there, answering when she pleases, unrushed, rarely smiling but never frowning, her handsome, smooth-skinned face a perfect mask for whatever machinations click and whirr behind it.

Dragovan was raised in British Columbia's Okanagan valley. Her father was Slovakian, her mother Irish, and it is tempting to look for some mystical, genetic impetus that has thrust Dragovan into the rough and play-for-keeps universe inhabited by the mining promoters of Vancouver. There is the matter of her father. When still a bachelor, the young immigrant headed for the goldfields of northern Ontario, where he worked underground until the Great Depression drove him west.

"Don't look for a reason there," says Dragovan. "Don't think I got the gold fever from my father. When he worked in the gold mines he operated a piece of equipment so dangerous they called it the 'widow-maker.' No, there was no great passion for gold handed down from my father."

In fact, what Dragovan wanted to be in life was a teacher. An aunt she admired was a teacher, and that is where the idea was born. But her family pushed her towards teaching, and pushing Nell Dragovan is bound to achieve the exact opposite of what is intended.

"I just decided right then that if there was one thing in life I would never be, it was a teacher."

Dragovan drifted into the law firm after university in Vancouver, but once she made up her mind that there was nothing separating her and the clients but some clever positioning and a whole lot of money, she didn't waste any time. From legal secretary, Dragovan went to work with friends on a market tip sheet. Almost as soon as she arrived, the friends left for other ventures, and Dragovan found herself the sole proprietor, editor, reporter, proofreader and sales agent. She learned a lot of lessons, fast, and got out.

And so to real work.

Dragovan went to work with the promoter Sankar Ramania. Ramania was, and is, a very successful promoter, and he had run up a couple of rather rewarding properties in the petroleum line of country. Under Ramania's tutelage, but nonetheless entirely on her own, Nell Dragovan took the plunge into promoting.

First, she formed her own company, Tri-Star Resources. Tri-Star owned a little mining property up in the Kamloops region of the British Columbia Interior. There was a little exploration work done on the property, and a prospectus was issued.

A prospectus is a statement of material fact that a promoter must prepare before going out to beat the bushes for shareholders. You are supposed to detail the property; assess its worth; record its performance, if any, and state the number of shares you are issuing and for how much. You are also supposed to tell the prospective purchaser of your shares what you intend the company to do with the money obtained from investors. There are a whole batch of other legal minutiae that you must observe when you write up a prospectus, but that's it in shorthand.

Then you hit the streets and hawk your paper.

In Vancouver you must sell a minimum of 250,000 shares if you are to get your stock listed on the exchange, and you must make those sales in four months. This doesn't sound so hard, but it is. In Vancouver the streets are choked with smiling youngsters in new suits. All of them carry far-too-expensive attaché cases with exotic airline tags dangling from the handles. It is as if they had just flown in from Switzerland but — hey! — somehow have the time to fit you into their lives. Those briefcases, unmarked by the dust of any streets but Vancouver's, will be chock-a-block with fresh paper hot off the press.

Add to this lowing herd of anxious tyros the real pros, the bucket shop barkers with a dozen guys on the blower trying to hype and wheedle and bluster their stock onto the market, and you get some idea of the sweating world into which Nell Dragovan stepped with, ahem, Tri-Star Resources.

Dragovan is a tall woman, and striding along the street from one of her high-rise hiding places to another, nodding at this acquaintance or that, dressed to kill and looking like a million bucks, she is the picture of success. To see Nell Dragovan in her milieu today, comfortable, calm, authoritative, slicing whole passages out of agreements and sending them back to the lawyers with curt notes — seeing this it is hard to imagine she ever had to struggle. But she did.

"It was hard slogging for those four months. Boy was it hard!

Believe it. Those shares were priced at twenty-five cents apiece, but they were still hard to sell.

"Picture it. I walk into somebody's office to try to sell him stock. He asks who the directors of the company are. I tell him that I'm one of the principal directors. He asks me what experience I have. I tell him none." Dragovan doesn't smile, but she looks a little whimsical for a moment. "Well, would you buy it?"

But before Tri-Star was even listed, Dragovan got a break. Bob Lamond, a Calgary high-roller, was looking for a resource company listed on the Vancouver Stock Exchange. He wanted control of a Vancouver company and was looking for something he could buy right away. He could buy Tri-Star right away, Dragovan told him, and he did.

"That didn't hurt my reputation, that sale. The stock went up to a high of $7 on the exchange."

Perhaps the barest flicker of a smile.

"I made a profit."

It is inevitable that someone who starts a company — her *first* company — whose stock takes the elevator from the basement at a quarter a share to the penthouse at $7, will attract attention. Nell Dragovan attracted the attention of no less astute a market personality than Murray Pezim, the eagle among the hawks, and Pezim invited her to join him in a loose association. The idea is that if you get enough clever people braining away in the same suite of offices, *every*body is going to get richer. And so, trusting Nell Dragovan's judgement, Pezim steered Steve Snelgrove and the Hemlo claims her way.

Dragovan liked what she saw, and struck a deal practically on the spot. She happened to have a company on hand at the time into which, she judged, the Hemlo property would fit very nicely. That company was called Corona Resources, later International Corona.

It would be wrong to say that Corona was Dragovan's company, however. It was really no one's company at that time. Corona Resources was just a washed-out shell of a company, one of a spawn of companies formed by Dragovan in the heady days following her success with Tri-Star. But they needed *some*where to park the claims, and Corona looked as good as anything else.

Pezim made Dragovan the president of the company, and they were in business. For Dragovan, being made the president of a company was about as exciting as flipping for the coffee. This was strictly routine, the daily work of a promoter. "To me, Hemlo was just another property. There was nothing special about it. There had been some work done on it and there had been some gold showing, but so what? There are lots of properties around with those characteristics. And they're not all sitting on billion-dollar ore deposits.''

What really convinced Dragovan and Pezim to get behind the Hemlo property with hard cash was the enthusiasm of Dave Bell. Acting on the advice, perhaps the insistence, of Don Moore, Steve Snelgrove had strongly suggested that Corona hire Bell to manage the drilling program necessary to assess the Hemlo claims.

''Dave Bell took a very ballsy approach to that property,'' Don Moore says. ''And when it got to her, Nell Dragovan stuck *her* neck out.'' Moore is referring to a cheque for $10,000 that Dragovan gave to Bell when he wanted to get going with his drilling program in advance of the underwriting that would raise the money for Corona's treasury. ''Nell wrote that $10,000 cheque out on her own account. I remember her telling me that when she sent that cheque out, she thought 'What am I doing?' ''

Recalling the incident herself, Dragovan says, ''I decided, well, I'll send David $10,000 of my own money. If he steals it, it'll be a $10,000 experience.''

But David Bell didn't steal Nell Dragovan's money. David Bell went drilling with it. And David Bell found one brain-busting great whacking rock full of gold.

So Nell Dragovan struck one of those rare deals that does the magic thing: a promoter likes a property; a promoter buys a property; a promoter promotes that property; that property gets developed; that property becomes a gold mine.

It doesn't happen often, but when it does, it is sweet to be a part of it.

Here is the deal Nell Dragovan and Steve Snelgrove cut: Corona would pay $40,000 in cash right away, and 300,000 shares of Corona stock. Naturally, if the property turned out to be a producing gold mine, the stock would be worth a fortune. The stock was to be issued to the partners in increments of 50,000 shares, spread out over exploration and drilling, with the final installment due when the mine went into production.

It is important to note here that the partners at that point in the development of Hemlo included Steve Snelgrove. When Don McKinnon and John Larche made their deal with Snelgrove, the deal by which Snelgrove would attempt to market the claims, the deal was that by finding a buyer — in effect, a corporate partner — Snelgrove would earn 50 per cent of the property.

And so of the first $40,000, $20,000 went to Snelgrove, and of the first 100,000 shares, 50,000 were Snelgrove's. So far, so good. McKinnon and Larche dive back into the bush, slamming staking posts into the ground around the original block of claims as fast as they can. Steve Snelgrove? Well, Steve heads back to Bay Street with the stock in his pocket. *All* the stock. Too bad. Because through a number of unfortunate events the shares were sold to cover losses incurred by Snelgrove in margin trading; and whoosh went the whole block straight out onto the street with a new owner.

There was no trouble finding buyers. By the time this took place, Murray Pezim had been working the Corona stock with a good deal of gusto and it was trading at some very promising levels. When Don McKinnon and John Larche found out what had happened, they were more than a little discountenanced. In fact, Larche would have liked to drive his backhoe down to Toronto and scrape the overburden off the top of Steve Snelgrove's head.

''We did what we could,'' Larche remembers. ''I had a lawyer draw up an agreement, and I confronted him with it. The deal was, he was to sign over everything, any stock he had coming, and his royalty.''

There are two kinds of revenue interest that the original participants in a staking can retain when they deal off their property to a company more able to raise the tremendous capital needed for detailed exploration and ultimate development. They can retain a share of the net profits, or they can take a royalty on the production coming out of the smelter. Which is better depends upon the circumstances of the day.

If, for instance, the price of gold is particularly low, then naturally the profits of a gold mine will be reduced and the holder of an interest calculated on the net profit may find himself with sharply reduced earnings. In the case of a net smelting royalty, on the other hand, the royalty does not depend on whether the mine is making a profit or suffering a loss. The payment is calculated on the gross value of the bricks coming out of the mill, with a few minor deductions to cover security and transportation between the mill and the refinery. If the price of gold, and consequently profitability, is low, then a smelting royalty is the thing to have. But if the price of gold is high, then the net profit interest can rise higher, and faster, than the royalty.

There is another consideration. In a venture such as that which is currently developing The Golden Giant, the capital investment of the major must be repaid before the participants begin to share in the profits. This is not the case with a smelting royalty. With a smelting royalty, you must pay the person holding the royalty on every brick you produce, whether you have recouped your own investment or not.

"Steve didn't want anything to do with it. He said he'd look at it later, that he didn't have time right now. He said he didn't have the corporate seal with him. But I said, 'Steve — no more bullshit!' and he signed the agreement."

No one knew just what Steve Snelgrove was signing away that day, not even Larche.

"I had tried to get someone to put a value on that smelting royalty, but they couldn't. They wouldn't even try. We didn't know what was up there. It was just a deal. We'd lost about $500,000 apiece, Don and I, and we were just trying to salvage a deal the best way we could."

As it turned out, when he signed that agreement, Steve Snelgrove signed away something like $3,000 a day for the rest of his life. Because once they started to poke holes into the rock around Hemlo in earnest, why, they just kept hitting gold.

Four

The Grubstakers

Claude Bonhomme and Rocco Schiralli. These are not names that grace the scions of old, aristocratic families, true. And these are not names that you will find on the membership list of the Toronto Club. But what these names *do* signify are two very cagey, very bright and soon-to-be very rich men. For Claude Bonhomme and Rocco Schiralli are two of the wiliest grubstakers in the trade, and a grubstaker, if he grubstakes the right venture, brings home the bacon in a Brink's truck.

A grubstaker is a financier on a modest level. What the grubstaker does is finance a prospector who wishes to stake a certain specified piece of ground. In exchange for the cash, the grubstaker takes a piece of the action. If there is any. It is important to remember, in a chronicle like the Hemlo story, that a prospector does not come into an office with a piece of paper on which is written the sum of $5 billion. A prospector comes into the office only with a spiel and a chestful of hope. It is here that the grubstaker must make his call, discerning the nugget amidst the densely flying bullshit. If he calls aright, party time. If he draws a dud, it's good-by to the bucks.

Our meeting is set for eight o'clock in the morning. If you want to catch Bonhomme and Schiralli, go for the early appointment. The rest of the day is devoted to dealing with people with more interesting commodities than mere questions. Bonhomme arrives on time, a jolly, short man buried under one of those towering wedge hats for which some glossy little beast of the forest has paid with its life. He carries a bag loaded with great tubs of coffee from the downstairs take-out joint and leads the way into his office.

The building that houses Bonhomme's office is at the corner of University Avenue and Adelaide Street, no more than a block from the Toronto Stock Exchange. More important than that, it is joined by an arcade to the sumptuous lobby of the Westin Hotel. It is here that Don McKinnon and John Larche stay when they are in Toronto, and it is here that other people who want to talk to Bonhomme and Schiralli stay. Convenient. Fast. No time wasted.

Bonhomme's office faces University Avenue. The room is spacious, and appointed to Bonhomme's peculiar taste in a profusion of warm, glowing marble. The coffee table, boardroom table and the broad surface of the desk: all are fashioned from single, thick slabs of pale marble supported by wrought iron. Bonhomme is wearing a blue blazer, a shirt and tie and a lively sweater bearing a diamond pattern. His trousers are borrowed from a blue pin-stripe suit. Dark brown eyes watch you from a face that is both animated and controlled, ready to wrinkle into a smile, but ready for business, too.

In 1960 Bonhomme abandoned a faltering construction business in Sudbury and moved north to Timmins. He met Don McKinnon the next year, when McKinnon was sitting as a member of the local industrial commission and Bonhomme was immersing himself in an uncle's thriving real estate business. Over the years the friendship developed, with Bonhomme selling the prospector new houses and buying his old ones. "I even ran against Don for a seat on the Timmins council one year. He won."

It was not until 1969 that Bonhomme made the move into the mining business. "A group of us formed a grubstaking syndicate. There were fifteen members, all from the Timmins area. We put together two companies, but we didn't have anyone to run them. So I became the president. That was my initiation into the mining business."

To find out what the staking business was all about, Bonhomme bought himself a prospector's licence and headed into the northern bush. He pounded in a little wood, scrawled his name and licence number on a few posts, and registered some claims. But the expertise didn't seem to help. "The four years from 1973 to 1977 were terrible years in mining. They were terrible years for promoters. That's when I got into oil and gas properties."

Rocco Schiralli sweeps into the room, twenty minutes late. Schiralli is the fashion plate of the partnership. Today he is robed in a long, black leather coat, lined with thick fur, that drops almost to his ankles. Shucking off this splendid garment, Schiralli emerges in an impeccable dark blue suit, gleaming white shirt and discreet tie. His shining black loafers glitter with tiny golden buckles as he crosses his legs and settles into an armchair.

That the two partners are also friends is obvious from the start. There is a lot of kidding about punctuality, and a great feeling of camaraderie in the room as little paper packets of sugar get tossed back and forth and the lids are prised off the coffees.

"It was early in 1980 that Don first came to see me with those Hemlo claims," Claude Bonhomme says, stirring a little cream into his coffee and glancing over at his partner. Schiralli confirms the date with a nod, and Bonhomme continues.

"In 1980, January or February, and he wanted $50,000 for the fourteen claims. Those were the claims that Corona eventually took. I said I didn't have $50,000, but I'd look around. I approached the majors and I approached Schiralli. Rocco was my solicitor at the time, and he had clients who were always looking for companies."

Schiralli nods and gets to his feet. He appears to be more comfortable pacing a little as he talks, gesturing and commanding his own space. "My clients were always in the oil and gas business, but when you're trying to raise money you look at everyone you know. Still, I didn't approach clients on those first claims — the Corona claims.

"Claude was talking to me — he was bugging me, really — trying to get me to put up $50,000. I'll never forget it. I was standing in the line-up at my bank on a Friday afternoon. I was just going to get out a couple of hundred bucks for the weekend. All of a sudden Bonhomme here comes walking in, and stops. He shouts over to me, 'Hey, Schiralli! Are you going to take those claims?' I shout back, 'How much?' Bonhomme yells, '$50,000!' So I yelled back, 'No! Sell them to someone else.' "

Ultimately, they were sold to someone else, but luckily for Bonhomme and Schiralli, Don McKinnon was not through pounding on their door. He came back to see Bonhomme again. "He told me that they were doing a deal with someone else on the original 14 claims. But he said, 'Look, there's eight miles of strike [mineral occurrences] on both sides of those claims.' So I asked him how much it would cost to stake and he told me $100 a claim. I asked him how many claims he figured he should stake, and he told me 150. Okay, that was $15,000. I went back to Rocco."

"I hadn't even seen a map of the place," Schiralli says. "I didn't even know where Hemlo was. But I said, 'We'll form a grubstake.' I knew I could rely on Claude's word, so I asked him if he thought it was worth it, and he said yes. So I said 'Okay, let's go.' "

Schiralli drew up the grubstake agreement immediately, simply dictating the terms to a stenographer. "When you've been doing deals like this as long as I have, you carry the details around in your head. I don't have to consult any references to dictate a grubstake deal." Under the terms of the deal, Schiralli and Bonhomme's wife, Helene, formed the grubstaking syndicate, and McKinnon and Larche were the prospectors. Schiralli and Helene Bonhomme would share a two-thirds interest in the claims, and Larche and McKinnon would hold the remaining one-third. The deal was signed on 9 June 1980, and Schiralli handed over the first installment of $5,000.

In fact, *all* the money for the grubstake came from Schiralli, even though he and Helene Bonhomme shared equally in the syndicate. The money put up by Helene Bonhomme originated in a loan from Schiralli. Mrs. Bonhomme should have no trouble paying back the loan. The claims that Don McKinnon and John Larche went out and staked now comprise The Golden Giant deposit, with reserves as of January 1985 of 24 million metric tonnes grading 0.28 ounces of gold per ton. That is 6.72 million ounces of gold which, even at the miserable prices current in early 1985, was worth more than $2 billion.

About the time that Schiralli was handing over the last cheque for $5,000 — August 1980 — the prospectors' deal with Steve Snelgrove was falling apart. Snelgrove was begging McKinnon and Larche for more time to come up with financing for the original fourteen claims. The prospectors agreed, and Snelgrove finally took the claims to Vancouver and struck the Corona deal with Murray Pezim and Nell Dragovan.

"Our own claims were actually acquired *before* the Corona deal," Schiralli notes, proud that he and his partner scooped the people credited with breaking the Hemlo gold play. Bonhomme agrees, and adds, "Now we had to start doing the whole thing all over again. We had to start knocking on doors."

The door-knocking was loud enough to catch the attention of a pair of astute mining promoters headquartered in Vancouver, and early in 1981 Richard Hughes blew into town ready to buy a gold mine. Hughes and Frank Lang, his partner, are two of the most respected mining entrepreneurs in the country. From its offices atop the old

Royal Bank building on Vancouver's West Hastings Street, the Hughes-Lang Group controls as tight and successful a knot of junior companies as any promoter could wish for.

When Hughes flew into Toronto, he headed straight for the Westin Hotel, where Rocco Schiralli and Claude Bonhomme joined him. Early one morning the partners met in their offices, assembled their papers and trooped through the arcade into the lobby of the Westin to meet Dick Hughes for breakfast.

"We wanted to do business with him because he was associated with a producing mine," Schiralli says. "Hughes and Lang already had a gold mine in Quebec that they were operating themselves, and that made them attractive people to do business with."

They negotiated a deal with Hughes that day. Hughes returned to Vancouver and had his lawyers draft an agreement putting the claims into two Hughes-Lang companies, Golden Sceptre Resources and Goliath Gold Mines. The agreement arrived back in Toronto, but Bonhomme didn't like it. The partners came in on a Saturday and redid the whole thing. "We did it fast," says Schiralli. "We used their agreement as a starting point, and in an hour we had it done. No longer. And you know, looking back on that agreement now, if I had to change one thing in it, I don't know what I'd change."

The agreement was sent back to Vancouver, and accepted. Under the terms of the deal, Hughes and Lang paid the partners $25,000 up front with $75,000 to come in staggered payments as development progressed. For their money, Hughes and Lang acquired an 85 per cent interest in 156 claims that would turn into the fabulous Golden Giant deposit. The grubstakers and prospectors retained a 15 per cent interest in net profits.

Our second meeting is at Rocco Schiralli's law firm, Armstrong and Schiralli, just along the corridor and around the corner from Bonhomme's, on the other side of the bank of elevators that is already busily shuttling the tenants into their starting blocks for another day. Like most downtown law firms, Schiralli's has an opulent reception area, all softly glowing panels and handsome sofas. Along a corridor, Schiralli's own office is more modest. Few businessmen with their priorities screwed on right will surround themselves with the same lavish furnishings into which they plant their receptionists. This is not because they cannot afford it. If he felt like it, Rocco Schiralli could certainly afford an office big enough to play tennis in. But Rocco does not come downtown to bat fuzzy balls around. He comes here to work.

Rocco Schiralli's success story is the classic tale of the North American dream. Unlike his partner Claude Bonhomme, whose family comes from generations of that sturdy, tough French-Canadian stock that peoples northern Ontario like a species of indestructible tree, Schiralli's is the immigrant story. The eldest son of a Canadian mother of Sicilian descent and an Italian immigrant father, Schiralli's childhood and youth were all passed in Toronto, where his father ran his own corner grocery.

Schiralli's first love was engineering, and in 1962 he graduated from the University of Toronto with a degree in engineering physics and thermodynamics. He went to work for Republic Steel in Buffalo, New York, as a combustion engineer. "That's what I really wanted to do. I wanted to work in gas dynamics. But you know, I just couldn't keep my mouth shut. I was too candid. People would ask me what I thought of something, and I would tell them."

Schiralli lasted only one year at Republic. He returned to Toronto to work for Consumer's Gas, programming a customer accounting system, until he decided to return to university and study law. "I used to think that lawyers were just a bunch of crooks, while engineering was the noble profession, working with the great laws of nature. I don't feel that way about law any more. I feel that without the law there would be nothing to protect the individual from the state or from other individuals.

"Besides, I was always getting into difficulties with senior management when I was an engineer. I was not a very good employee from the point of view of being docile."

Schiralli lives with his wife and three children in a building where the condos start at $300,000. On weekends he takes the family north to his lavish spread on Peninsula Lake near Huntsville, a three-hour drive from Toronto. There he shares the shoreline with such landowners as the proprietors of Deerhurst Inn, a super-luxurious watering hole and sports playground settled into some of the most beautiful scenery in Ontario. Schiralli's palace used to belong to Tim Horton, the hockey player. The living room is vast, almost seven hundred feet square; the cathedral ceiling is twenty-five feet high. Tall windows frame the forest and the lake.

"If you live at the corner of Bloor Street and Bay, where there's no grass to cut, no painting, the doorman gets your car and parks it, there's not even a balcony — then you have to have the opportunity to do *something* outside. So that's where I go to cut grass and hammer nails."

There is an immediate burst of warmth when Claude Bonhomme enters the room. There is something so genuine and so full of optimism in his energetic deportment that it is contagious. When you speak he watches you closely, ready to laugh if you want to laugh or expound with patience if you are puzzled. He must have been a popular teacher before he abandoned his Sudbury classroom back in the 1950s and headed for the uranium boomtown of Elliot Lake to sell prefab houses to the clamouring populace.

Bonhomme drifts over to the elaborate, multicoloured map spread along one whole wall. ''That's the Abitibi Gold Belt,'' he says, moving closer to peer at the map. ''It runs from northern Quebec into northern Ontario, and most of Canada's gold production comes out of that belt. There are three thousand established gold deposits in the Abitibi belt, each one marked, each one mapped.''

Bonhomme moves around the long table that is buried under piles of legal papers, and taps at the plastic covering of the map. ''This is essential to us, this map. If somebody comes in to sell Rocco and me a property, we can use this map to find out exactly what work has been done on that property before, who owned it, what the showings were — everything we need to know.''

Schiralli talks about a reorganization of the original grubstaking syndicate. ''In August of 1981 we decided to roll over the interest in Hemlo to a company we called Hemglo Resources Limited. The name Hemlo didn't appeal to me, so I just added a *g* to make it shine.

''The problem with four people operating under a grubstake agreement is that it's like an association — there are no rules. You have to agree on everything together; there is no framework in which to act except by absolute agreement. By forming a corporation, we had flexibility. We could speak with one voice.''

He mentions one occasion when the single voice spoke for all, in the spring of 1982. Results were starting to come in on The Golden Giant drilling program. And those results were looking good. At that point, Claude Bonhomme was monitoring his claims meticulously, and he spotted one about to come open. The paperwork needed to maintain the right to the claim had simply not been carried out and filed with the mining recorder in Thunder Bay. The claim would shortly become available for anyone to stake. The duty to ensure the legal maintenance of the claims lay with Hughes and Lang, the grubstakers maintained, and they took swift action.

Schiralli wrote to the mining commissioner and asked for an extension of their rights to the claim, which they got. He then wrote to

Hughes and Lang in Vancouver notifying them that they were in default of the terms of their contract, and giving them sixty days to comply. There was no response. When the sixty days were up, Rocco Schiralli wrote saying the agreement was terminated. That time there was a response. A subsequent agreement restored the original, and there are now measures in place to ensure that all the assessment requirements needed to hold the claims are performed at least six months before they are due. When there are several billion dollars' worth of gold at stake, you don't want some stranger banging fresh wood into the middle of your property and telling you to move over.

Claude Bonhomme has left half an hour ago. Rocco Schiralli is starting to glance at his watch, and his secretary is hovering in the doorway with a pile of documents that looks thick enough to handle the transfer of the Panama Canal. There is moose pasture out there waiting to be staked, by God, and as long as there's a chance there might be gold underneath it, stand aside. The grubstakers are ready to do business.

Five

The Jewish Howard Hughes

Joey Bishop, the famous comedian, is talking. This is a man who pals around with Frank Sinatra, Sammy Davis, Jr., Jonathan Winters, Phyllis Diller, Buddy Hackett and about a hundred other stars. But today Joey Bishop is not talking about any of these stellar folk, some of whom he has known for thirty years. No, today he is talking about a Canadian financier and stock promoter, his old friend, his dear friend, Murray Pezim. Bishop is sitting in his good buddy's Vancouver office.

"You know something about this guy? This guy is the only man I know who could walk down the street with a million bucks in each pocket, eating a smoked meat sandwich and wearing the mustard all over his lip."

Ha ha ha. Everybody laughs, including Murray Pezim. Murray Pezim laughs because the purpose of the observation is to show what a regular guy old Murray is. Murray is just folks, Bishop seems to be saying. That is why Murray and the faithful retinue attending him at this moment all laugh. They laugh because they know it suits Pezim to be known, at least right now, as good old Murray — hey! a lovable guy! — the millionaire who's so gosh darn down-to-earth that he goes around with a couple million bucks in his pockets and mustard all over his big, friendly puss.

But really, while everything that Bishop says about his friend is true, none of it is all that funny.

Certainly, Murray Pezim could walk down the street with a million dollars in each pocket. It would take Murray Pezim about fifteen seconds in the bank to emerge with two bundles of a million dollars each, and the bank manager would probably be following him to the door to make sure it was enough. It is a great thing to be able to barge into your local bank and scoop a few million out of the till in less time than it takes most people to find the right line, but it is not a *funny* thing.

And, sure, Murray Pezim is the kind of man who could easily be found walking down the street eating a smoked meat sandwich and, okay, there could be mustard on his mouth. But if Murray did choose to have his lunch walking down the street instead of having it delivered to him right at his desk while he worked, then there would be a very good reason for that. And if Murray had mustard on his face while he was walking and eating, that mustard would not be there because Murray is, shucks, such a big dopey fella that he plumb forgot it was there. No, sir. It would be there because Murray Pezim had calculated that it was in his interest, at that precise second, to wear a little mustard for the grinning crowd.

"This guy," Joey Bishop is adding, "this guy is the Jewish Howard Hughes." True. And nobody ever suggested that Howard Hughes got where he got by being sloppy with his food.

There is a CBC documentary film on the Hemlo gold strike that contains a sequence showing Murray Pezim returning to his boyhood haunts. This is a journey that a lot of men make. They make it like a pilgrimage, on foot, walking up the old street and standing dumbfounded before the old house. They find the tree they used to shinny up when they were boys, and they note that the tree now has a diameter of two feet. They note with astonishment that some savage has torn the veranda off the front of the house, paved the lawn for his BMW and battered a lot of light into the old place, light that was never there before. They shake their heads and walk on, a little sorrowful.

Not Murray.

Murray sweeps up Toronto's Spadina Avenue in the back of a white Lincoln limousine. The car is so long that it looks as if there might be a bowling alley inside. There is not. There is a small television set, and of course there is a phone. If there were no such invention as the mobile phone, Murray Pezim would probably have to maintain a whole crew of linemen, following him around stringing telephone wires behind him. You get the idea: this is a man who likes to stay in touch.

The limousine whispers up the busy thoroughfare, turns west onto Baldwin Street and drifts into the mad, whirling chaos of Kensington market on a Saturday morning. Murray has come home for a look-see. The shoppers jamming the sidewalks and spilling out on the street try to see in through the smoked glass. Murray is peering out, remembering.

''Yeah, these guys are all Portugese in here now, but this used to be the Jewish ghetto. It wasn't very fashionable to be a Jew in those days; it was pretty hard. The school was only a couple of blocks away, but to get to it you had to fight your way through a lot of tough kids.''

Pezim chuckles to himself as he stares out the window at the curious shoppers. ''Well, you didn't *have* to fight. You could walk around, go maybe ten blocks farther in a wide circle.'' He laughs again.

''But that was too far. So I fought. That's how I got this nose.''

Pezim wears a big grin as he looks at you and fingers what is in fact a pretty impressive appendage. This is not so much a nose here as it is a piece of sculpture. It is sculpture in that it represents an ideal that the proprietor wants the world to know he subscribes to. That ideal is that you never, ever give up. It is true that a lot of savage little sculptors used to lie in wait around the fringes of the Jewish ghetto in Toronto in order to have a chance to take a swipe at that particular work, but it was Murray Pezim who was directing the overall project. The nose is a monument to Murray's determination that nobody was going to decide which route Murray took to school but Murray. And if now he honks and blasts and echoes through a set of somewhat battered tubes because of it, well so what?

The limo is having trouble negotiating the crowded streets this morning. This is not Rosedale in here, and the boiling confusion of a Portuguese fish market is no place for a quarter-mile of car. But although the morning mob is breaking around his shores like surf, The Pez is elsewhere. He is lost in contemplation as he studies the tops of the old houses which appear above the remodelled store fronts that line the street. This is a little like trying to pick your grandfather out of a line-up of foreheads, but The Pez is squinting hard. Finally he sits back in the seat and nods, tapping the window with a polished finger.

''There it is. Jesus. That's where I grew up, boys.''

There are murmurs of wonder from the jump seats as several of The

Pez's retainers stare out of the car and crane their necks for a glimpse of Bethlehem.

"Up there on the top, the left-hand window. That's my room up there. The butcher shop was downstairs."

In 1945 The Pez went into that butcher shop, rolled up his sleeves and started to work like a madman. He cut, he wrapped, he delivered. He sold, he sold, he sold. Pezim had no very definite idea what he wanted to do with his life back then, except for one central purpose: he wanted it to be better. He wanted to be a success, and although he may not have known exactly what he meant by success at the time, he was fairly certain that he'd recognize it once he got there.

By 1949 Murray had saved $13,000. Thirteen grand represented a substantial stash in 1949. You could buy yourself one hell of a house back then for $13,000. If your salary had been $13,000 in 1949, you would not have been one of Murray Pezim's customers. You would have been ordering your steaks from whomever it was did the fetch-and-carry trade in Rosedale, and that would not have been Murray. Yes, it was a lot of money to anybody. To the people stuffed into that tangle of bursting streets west of Spadina where Murray lived, the $13,000 represented a miracle of thrift and grit.

The Pez blew the whole stash in six weeks.

The Pez went downtown and took Lesson Number One in the market, and he had some very tough teachers.

There were lessons that Murray Pezim learned in those six weeks that have stood him in very good stead for the last forty years. He learned that people who dress in expensive clothing and speak in soft, starchy accents, people who are chauffeured home to Forest Hill in time to give the dog a bath, are not always nice people. Sometimes, he learned, they have spent the day giving someone else a bath.

Murray learned that you do not have to scream and shout and pound the counter with your fist in order to make a deal. He saw that one marketplace was not so very different from another, and that traders still haggled and cajoled and bargained and threatened, but that it was possible to do all this with perfect teeth and a membership in the Toronto Club. Yes, some extremely well connected folk taught Murray these lessons, and some of them, you can bet the farm on it, wish to hell they hadn't. Because Murray is paying some of them back in spades today or, if not them, their sons. It needn't have been that way, but it was. By insisting upon relieving the brash young butcher from the Jewish ghetto of every last red cent he possessed, they led him to the burning, paramount conclusion that they would all have been better off not leading him to: "I made up my mind right then that any business where you could lose $13,000 in six weeks, I wanted to be in that business."

And now he is, by george. And if anybody is laughing down there on Bay Street, then somebody better tie him up and cart him off to the insane asylum. Because the butcher they peeled thirteen grand off of in 1949 has come back to their front yards and walked away with the richest gold mine in the Americas. If anybody is laughing now, boys, it's the guy with the beat-up nose.

What The Pez did was simply march his thoroughly kicked ass right down to Bay Street and hunker down. He started to haunt the brokerage offices. With no very definite idea of what he was doing, The Pez would sit there and stare at the board as the frantic kids dashed to and fro chalking up the bids. Just sitting there and staring at the board is a good way to avoid attention on Bay Street, particularly if you have no very definite idea of what you are doing. Clueless board starers are as populous on the street as pigeons, and just about as important to the financial life of the nation. Nevertheless, what The Pez saw sparked him enough to hightail it down to New York, where he managed to get a job working for a Wall Street broker — without pay — for six months.

Returning home to Toronto, The Pez parlayed what he had picked up in New York into a job as a salesman for E. T. Lynch and Company. He lasted there for a year and a half and moved over to Jenkins, Evans and Company, a firm that no longer exists. But it was a company on the boil in those days, doing a land-office trade in promoting penny mining stocks. Whatever it was that he smelled cooking in the kitchens of Jenkins, Evans, The Pez decided it was his kind of food. He stayed with the firm for seventeen years.

To say that Murray Pezim prospered is to say the sea is wet. Within five years the butcher from the Kensington market made himself a rich man, and the thirteen grand he'd dropped for his first lesson in market ethics was beginning to look like a shrewd investment. Enter Stephen Roman.

Roman was a young promoter, born in Czechoslovakia, who was looking for backing. He was trying to raise $40,000 to finance a drilling program on a uranium prospect near Blind River, Ontario. The Pez helped Roman finance the drilling, and the rest is history. Roman struck a fabulous uranium deposit at Elliot Lake, and the company he controlled grew into a cash machine.

"Well, you know the rest of the story," Pezim recalls. "It was Denison Mines and it was trading at forty cents a share. My people and I made fortunes, even though I sold my stock at $17 and then watched it go to $85 a share."

The Pez decided that a few rewards were in order, and he hopped a plane to Jamaica. After a little looking, he found a magnificent twenty-two-acre estate near Ocho Rios. He was looking for a place large enough to entertain in, and with three houses on the property this one looked about right. Pezim slapped down $497,000, moved in and started to send out invitations. The party lasted for a year and a half.

If you want to find The Pez today, you find him, when he's at home, in Vancouver, at 609 Granville. Pezim's offices are on three floors of the Stock Exchange Tower, squirrelled away here and there behind doors that are kept shut. If you want to do business there, be prepared to state your case with a certain amount of force and as much clarity of purpose as you can muster. If you are vague or circumspect or even too elaborately courteous, you will find yourself invited to take a seat, and someone will get you a cup of coffee while you wait. You will drink a lot of coffee up there on the fifteenth floor before you so much as catch a glimpse of Murray Pezim, if you take that seat. You will be grateful that the seat is comfortable, if you take it, because that is where you will be spending the entire day. Just sitting there, drinking coffee, and listening to the human traffic bellowing up and down the corridors beyond. The thing to do is *decline* the seat. Eventually they will send for Susan, and you are halfway there.

Susan is a ravishing woman, and also a smart one. If she were not smart, she would last only about four seconds as secretary to Murray Pezim, and then she would have to leave, probably in tears. Being good-looking is great, but it is not enough. The Pez does not suffer fools.

Today Susan is wearing a dress of astonishing elegance, reddish brown and cut low. Her jewellery is expensive enough that the only response it provokes up here on the fifteenth floor is guarded calculation. You can hear the numbers shuttling around in the brains as visitors to the office steal glances at the ornaments and at the peerless Susan.

Susan is tall and blonde and built by the same firm that did Esther Williams. There has been no stinting on materials. But Susan is no eye-batter. When she strides into the reception area to ask you what your business is, you had better know what you have in mind before she asks. Try to do much figuring out while Susan is standing there watching you with her beautiful, unblinking eyes, and the next thing you will see of the Pezim empire is the retreating back of his secretary. Tell Susan what you want; tell her it needn't take long; look as earnest as you can. If she smiles and says she'll see what she can do, you are home.

It is 6:30 in the morning. There is not much going on outside in the streets of Vancouver, but inside the building at 609 Granville people are so feverishly at work that you can almost see the smoke coming out from under the doors. The reason is the market. The market opens in Toronto and New York at 10:00. That's 7:00 in Vancouver and that means everybody is ready to cook by 7:00. No gang of pantywaste eastern bozos is going to get the jump on this shipful of pirates, by God, and the man who thinks he can show up to work for Murray Pezim later than 6:30 in the morning had better be ready to think about career options by 6:35.

When you are admitted through the door behind the receptionist, you are admitted into a world that bubbles and bakes with energy. All along one hallway are the offices of people associated with the Pezim operations.

"I don't want people working *for* me," says Murray. "I want people who work *with* me. If you can't work for yourself, then why should I be paying you?"

No one pays much attention to you as you follow Susan through the office. There is not a lot of idleness in these premises at 6:30. You could probably lead someone through dressed in a gorilla suit and they'd just think it was another of Murray's promotions. In one corner of the fifteenth floor there is a big open space, and a squad of busy stenographers sit at desks burning up the keys. Beyond the windows, away in the distance, the mountains of the Coastal Range march off along Indian Arm. It is late summer in 1984, and the clear sunshine is pouring down upon the beautiful city, kissing the water into a sparkle of light and filling the spaces between the mountains as if with some floating and luminescent sea. Everywhere you look in Vancouver the stunning prospects reach out to touch your eyes and beguile you from the dreary paces of the working day. But not on the fifteenth floor, 609 Granville. You could march the contents of the Vatican Museum past those windows and get nothing more adoring than a hurried glance. No one is there for the view.

There is a sign on the massive doorway that guards the inner precincts of Pezim's business. The sign says that unless you are one of the people who work beyond those doors, you are not welcome to come through them without an escort, who will announce you. It is a little forbidding if you have not yet met the man whose office it is. And while it's true that Murray Pezim makes friends very quickly, you'd better believe the sign means what it says.

When you walk through the heavy doors that guard the last perimeter, it is like walking into a cave. It is a comfortable cave, to be sure, but all of a sudden everything is a lot darker than it was in the outside office. This is because The Pez is not interested in the view. The Pez is not a landscape painter. The Pez comes in here every day at 6:30 to light the stove, and once he gets it lit, he wants to start cooking, not stare out the glass at a lot of big hills with trees on them. And so the curtains are drawn.

The office is about twenty feet wide and thirty feet long. The Pez needs this room. He needs this much air because otherwise he would die of anoxia. At any one time there may be as many as a dozen people entering, leaving, hovering, or shouting at one another in The Pez's office. If that office were not a big one, there would not be enough oxygen left over to fill the lungs of a titmouse, never mind the big bellows inside the chest of the boss.

There was a time when The Pez's office was even bigger. In those days, so the story goes, The Pez liked to keep all his trading staff in the same room with him. More cooks, more cooking. Then one day, in the matter of some rather tricky negotiations which were not proceeding precisely in the interests of one M. Pezim, The Pez lost his temper on the phone and tore a strip off somebody wide enough to slow down a pursuing army. No big deal. Happens every day with The Pez. Somebody makes The Pez mad; somebody gets a strip torn off. Maybe next time somebody is a little more careful. The trouble this time was that there were just too many ears listening to the strip-tearing, and within fifteen minutes there was nobody in Vancouver except the tourists who had not heard about The Pez's little problem with a certain party.

The next day The Pez had a smaller team in his office.

At one end of The Pez's office, just to the left along a wall as you enter, is a long and lavishly upholstered sofa covered in a pale pink velvety material. In front of the sofa is a large coffee table. Above and behind it, on the wall, are paintings. One is of a prize fighter; the other is an oriental work, or a work inspired by the Orient. It is a work of startling and arresting delicacy, a landscape that somehow commands the whole wall.

Seated on the sofa, his briefcase open before him and a phone in his lap, is a man behaving like a caricature of the street-savvy, slit-eyed New York businessman. As it turns out, this is what he is. His name is Shelly, and he keeps on bellowing at whomever will listen some incomprehensible gibberish about Bloomingdale's. Sometimes someone actually listens to him for a few seconds; and sometimes they give him a smile and an encouraging nod. Other times they just look away with a blank face, back to the screens.

The screens are everywhere.

In one corner beside the sofa, there is a giant screen six feet square. It is carrying financial bulletins and stock quotations, ceaselessly monitoring whatever markets Pezim wants to watch at the moment. The screen itself is about five feet off the floor, resting on a console that contains the sophisticated controls that make the massive unit do what its operators want it to do. It can play videotapes, record from someone else's tapes, record from the air, play back from other machines — can do just about everything but go get the coffee. And if there were not plenty of people around to handle the coffee end of things, The Pez would probably figure out a way to get the machine to do it.

Downfield from the sofa, at the closed-curtains end of the room, three massive desks are arranged in a U-formation, with the open end of the U facing out towards the room. The desks are made of rich, dark mahogany, and the tops are covered with exceptionally thick plates of glass. Atop each desk sit two of those complex, multibuttoned telephones and two screens. This makes a total of seven TV screens in the office. The whole place is enough like the control room of a television studio that you half expect someone to stick a head in the door and tell you it's ten minutes to air.

In addition to the two phone consoles and two monitors, each desk has a keyboard for punching up the call letters of any stock or warrant traded on the Vancouver Stock Exchange. You want to know what's going on on the trading floor twelve stories below? Tap-tap-tap, and there it is. The Pez sits at the desk in the middle, the desk that forms the bottom of the U. The drawn curtains are at his back. On The Pez's left sits Larry Pezim, a nephew. Larry wears a shirt and tie and looks exactly like a nephew working in his uncle's office, which is to say, not entirely at ease.

To The Pez's right sits Saul Cohen, an associate. Saul *does* look at ease. In fact, Saul looks as if he would be at ease in a tankful of sharks. Saul looks a little as if, were he to find himself in a tankful of sharks, those sharks who were not flattened up against the side of the tank farthest from Saul would be getting his coffee. This is not to say that Saul looks mean. He doesn't. There is just a certain clear-spoken severity about him that would encourage any creature with a predatory bent to give it a little more thought. Saul does not wear a shirt and tie. He prowls into the office dressed for southern California in a sweater virginally white, dark trousers and a pair of blue leather slip-ons with white soles. Saul's watch is so expensively discreet that it looks as if it could get into the Vancouver Club all by itself, which Saul probably could not. But this is just the decoration. What gives Saul his authority is his face.

The closest most people will ever get to a face like Saul Cohen's, unless they are sculptors, is when they stub their toes on a piece of granite. To contemplate the face of this man across the narrow width of his desk is to look straight into the definition of deadpan. And Saul's voice is to the monotone what the MX missile is to weaponry: simply the best available.

But Saul and Larry are only there because of the man who sits in the anchor slot. This is The Pez's show in here, and if the supporting cast is allowed a joke, or a spell of hoofing now and again, fine. Just let's all remember who the star is.

Today The Pez is wearing a yellow golf visor emblazoned with the words Pez Open. He is sponsoring a golf tournament to be held in Victoria, and somehow he has gotten it on the PGA tour on a one-occasion basis. If it is a success, if he can convince his fellow boosters in the business community to support it, then it will become an annual event.

In the central space contained by the U-formation of the desks, there are three vast chairs — great, sturdy monsters upholstered in crimson plush. The chairs look as if they would be just as much at home in the dressing room of King Edward VII as on the fifteenth floor of the Vancouver Stock Exchange. But who cares? They are comfortable, and so you just sink in, down to where your eyes are about level with the desk tops, and prepare to watch the action on the bridge.

Six

Just a Morning with The Pez

While the visitor from Toronto is off in a corner hearing from Joey Bishop what a great fellow Murray is, The Pez is holding two phones, one against each ear. If, say, he wants to talk into the one in the left hand, then the mouthpiece of the right-hand phone is simply swivelled out of the way. *Zap:* burst of words into the left-hand phone. Swivel. *Zap:* burst of words into the right-hand phone. Swivel. And so on. This is a difficult procedure to watch, never mind to master. All you have to go on is The Pez's end of things, although even by itself this is pretty impressive. The way it breaks down, the guy on the left, who is nameless, is getting a little encouragement from The Pez; the guy on the right, Paul, is not.

Right phone: ''No, Paul. No, Paul. That's not good enough, Paul, and you know it. That's bullshit, Paul.''

Left phone: ''Yeah. Ha ha. Way to go, buddy.''

Right phone: ''I'm not trying to tell you how to run your business, Paul. I'm telling you how to run *my* business. You understand me, buddy? And I'm telling you that's bullshit.''

Left phone: ''Yeah. Ha ha.''

Right phone: ''No, Paul. I got a call from one of the boys in the field, Paul, and you know what he told me? He told me he had to go into a stationery store and buy purchase order blanks to take those orders. *P. O. blanks* for Christ's sake, Paul. No, Paul. No.''

Left phone: "Yeah. Ha ha. Way to go, buddy."

Right phone: "That's right, Paul. You're finally getting it, buddy. Okay, buddy. Just go do it. Yeah, I know."

Left phone: "Yeah. Ha ha. Okay, see ya, buddy."

Right phone: "Okay, Paul. Go, buddy. No, Paul. Right. Go, buddy."

That Paul will now be out the door and about The Pez's business, few people in the room doubt. We got about eighty other businesses to run here, jeez. If Paul does not come around and start looking after The Pez's affairs with a little more alertness, then Paul's head will be rolling down Burrard Street towards the inlet with so little ado that his secretary will be piling stuff in his in-tray for days before she notices the strange, disturbing silence from the chair.

Like many promoters who weave their wondrous cloth upon the loom of the Vancouver Stock Exchange, The Pez uses many different threads. Shares and deals are shuttled around out here on the coast among a bewildering array of companies. Unless you've got a stack of paper businesses with your name on the directors' list, then who's to know how good you are? So there are plenty of gentlemen and ladies operating in and around the golden tower on Granville Street who have a company or two sitting in the top drawer of a desk waiting for someone to come along and park a deal in. But nobody like The Pez.

At last count The Pez had something like eighty-eight companies listed on the Vancouver board. These companies accounted for about 20 per cent of the total dollar volume traded on the exchange. Every time someone passed a buy or sell order across the floor to someone else, there was a good chance that some of the money represented by the trade was either going into or out of The Pez's pockets. In 1983 the Vancouver exchange traded about $330 million worth of stocks. The daily average was almost $16 million. That means that every single day of the year that traders were on the floor of the exchange screaming back and forth at one another, more than $3 million worth of the screaming was about the guy in the yellow golf visor sitting in the darkened room a dozen floors above their heads. This is called clout. This is fame. This is why there is a knock at the door, and one of The Pez's scuttling adjuncts announces that "the guy is here from the TV."

That is all that anyone in the room really needs to know. TV. An aide briefs The Pez on who the reporter is and what he wants to talk about, but The Pez does not really care, and neither does anyone else in the room. All that matters is that he is "the guy from the TV," and that he is going to give The Pez a chance to get the old mug on the tube again. This is just footage, the raw material out of which a promoter like Murray Pezim spins the magic which his reputation feeds on. It is this kind of accessibility that results in the instant recognition The Pez's name sparks. You mention Peter Brown and you will be presented with a face so blank you could type your resumé on it. And yet Peter Brown is the man most responsible for The Pez being here in the first place. Murray Pezim found the Vancouver Stock Exchange set up in a very agreeable fashion. Peter Brown, who now runs the investment firm Canarim, is one of the top insiders in the dizzying world of the Vancouver investment community. But who ever heard of Peter Brown?

Back on the fifteenth floor, Susan reaches down beside The Pez and pulls open the bottom drawer. There is a makeup kit in there, and she proceeds to brush on a layer of pancake over The Pez's already well tanned face.

"Believe me, it makes a big difference," says The Pez.

But no one else in the room is even listening. They have all been through this so often before that it is just another giraffe in the zoo as far as the keepers are concerned.

"Hey, Larry," Saul is calling across the room, "what's the symbol for Butler Mountain warrants?"

"B-M-T-dash-W," Larry yells back.

Larry answers his phone. The call has an immediate effect upon him, and he glances around the room quickly, noting who is present. There are a few people who are not exactly members of the family, and Larry swivels his high-backed leather chair around so that his back is to the room. There follows some urgent-sounding muttering. Larry puts the phone on his desk pad and walks around the desk to whisper something in The Pez's ear. The clowning stops immediately. The Pez pushes the makeup brush aside for a second, whispers a few short instructions, then rests his head back to let Susan finish. Larry returns to his own seat. He speaks into the phone in low tones:

"Buy 77,600, and hold them. Just hold them for us. Okay?"

The next call Larry takes has something to do with Swensen's Ice Cream. Swensen's is a company in which The Pez has a position. You do *not* have a position in a company if you own twenty-five shares that your grandfather bought for you when you were born, even if those shares have since split two-for-one and you now own fifty; shares are issued in the hundreds of thousands, and in the millions. Swensen's is an important enough company that The Pez had some promotional posters done up with him at a boardroom table, a couple of steely-eyed henchmen on either flank and some ice cream centre stage. It is behind Larry on the wall, right next to a spread of maps showing gold properties. So Swensen's is not a corner store. Thus Larry thinks it a not unworthy subject for The Pez's attention, and asks him something about it. Too bad for Larry.

"For Christ's sake, handle it yourself. I don't want to hear any more about the ice cream business. I'm out of the ice cream business right now. As of right now, the ice cream business is your business. Okay?"

Larry says it's okay.

"Good. I just don't want to hear another word about the ice cream business!"

Larry says okay, he won't.

Susan says, maybe the boss should put on one of his Pez golf shirts; the TV guy is here to talk about the golf tournament. There is an explosion of activity behind Murray's desk, and people seem to bloom into existence like flowers in time-lapse photography. Boxes pile into view and three or four people at a time are rooting through looking for the right shirt. Several are dragged into view until finally someone produces the right one. It is a white shirt, and it has the dates and other salient information about the Pez Open printed on the left breast. The shirt someone finally selects for Murray looks about a yard and a half wide.

"Hey, Murray, what size is that, a medium?" cracks Saul.

"Yeah. Ha ha," says Murray. The Pez's mind is elsewhere.

Modestly, Murray turns his back to the room and strips off the blue shirt. There is a lot *to* Murray. Susan helps him pull the white shirt back on over his head. Settled in the chair again, there are a few more slaps of powder onto the face where the shirt has rubbed it off. Murray lights a cigarette, puts the yellow golf visor back on, and looks like he's loaded for bear.

"Don't worry, buddy," he says to his Toronto visitor. "I'll tell you about my gold mine. You won't get any bullshit from me. Just soak up a little colour for a while. I gotta do this TV guy."

The cameraman is ushered into the room and begins to set up his gear.

"What station you from?" asks Saul, punching up numbers on his keyboard and watching the screen. Saul is being polite.

"CBC," says the cameraman.

"Right," says Saul, his mind elsewhere.

"You just do whatever you want, buddy. You just shoot whatever you want. Just tell us when you're ready,'' says Murray, reaching for a phone.

''Whenever the reporter gets here,'' says the cameraman.

''Sure,'' says Murray. The cameraman could ask the questions himself as far as anyone in this room is concerned. The reporter shows up, settles into his chair, and the interview begins. The reporter wants to know about the Pez Open.

''I love British Columbia,'' The Pez tells him. ''I think this would be a great thing for British Columbia, a great thing for the people. It would bring a lot of tourists, a lot of golf fans, into the province. It would be a great thing for us. I'm trying to get other businessmen involved, too.''

How does he find the time, asks the TV reporter.

''I love to work eighteen hours a day. That's all there is to it. Money, for me, money is just what you use to keep score. It's the *game* that I love. I'm not in love with money and I'm not in love with material things.''

The TV guy asks The Pez about the Mohammed Ali–George Chuvalo fight he tried to promote back in 1972. All of a sudden The Pez is paying close attention to the questions. He may not be in love with money, but nobody likes to get taken to the cleaners, either.

''I went broke. I got taken. It was my last quarter of a million and I lost it. There were a couple of guys up here from New York, and they knew a thing or two I didn't know. But I've got a good head on my shoulders, and I can always hustle a buck.''

Meanwhile the real business of the day is going ahead like a greyhound after a rabbit. The gentlemen are on the phones.

Saul: ''Anybody that has half a brain will be buying it now. It's a very tight hole. We're being very secret. Murray will disseminate the information very carefully. There's nothing to be afraid of there. Nothing.''

Larry: ''No, no. Don't sell it. Stupid people are selling it. Hold it.''

Murray: ''Okay, but I'm telling you, you got a day and a half, that's all. Okay, buddy?'' The Pez watches the screen for a moment, then grabs a phone. ''Scott, get those orders. It's not moving. No, no. Call and say, 'Why don't you buy? You know Murray.' '' And again: ''Believe me, Harold, you're my friend. I'm not looking for you on this one. This is real.'' And to Susan, who is having a difficult time sorting out what seems to be a very bitter complaint from a caller: ''Just hang up, honey. Don't give yourself a lot of aggravation. Just hang up.''

This is not just advice of the moment that The Pez is giving out here. This is operating philosophy. The Pez is famous for his techniques when it comes to extracting himself from phone callers who have become intractable. The Pez hangs up. Sometimes. Sometimes he takes a whistle which he keeps for this purpose, and blows it into the phone as hard as he can. The sound of a whistle being blown hard into a phone, and amplified by the ear piece at the other end will, The Pez feels, make the caller more circumspect about bothering The Pez again. This is what The Pez believes, and this is how The Pez encourages his staff to act.

"You want the whistle, honey?"

Susan contents herself with just hanging up, and you can almost hear the hiss of cold water on hot metal as the receiver is clicked down onto the shrill protests of the caller.

It is easy to forget, in the midst of the whirling promoterly buzz that fills this room, that the people on either hand are talking about real money. When The Pez and his cohorts bark their terse directions into these phones, real cash is changing hands, somehow, somewhere. This can have its ups, and this can have its downs.

In the early 1960s, when he was still on Bay Street, The Pez teamed up with another Toronto promoter named Earl Glick. Glickety-glick, they called Earl, who arrived in the world of high finance from the costume jewellery trade. The Pez and Glickety-glick undertook a number of adventures together, one of them an American company called Leland Publishing. The Pez remembers.

"The stock of Leland Publishing, at one time, was trading at $12 a share, and I had 300,000 of those shares. This is not exactly a lack of faith in the company, right? Then one morning I got a call from Leland's president, before the market opened. He said, 'Murray, the company's going into receivership, and I mean this morning.' All I could think of to say was, 'Can you wait until noon?' "

In 1964, when the Texasgulf copper strike near Timmins was driving the Toronto Stock Exchange crazy, The Pez and Glickety-glick were in up to their eyes. On a single day, when the TSE traded a volume of 28 million shares, 14 million, or *half the daily volume of the exchange,* was in stock controlled by Pezim and Glick. But a major stock scandal associated with the Texasgulf strike muddied the waters for some of the more energetic promoters as far as the Toronto Stock Exchange was concerned, and The Pez started doing deals in Vancouver.

One of his more successful early promotions from the growing Vancouver base was a company called Stampede Oil and Gas. Through drilling carried out by Stampede, Pezim actually discovered what turned out to be an important gas field in Alberta, the Strachan-Ricinus field. It was a major discovery, and it was totally accidental. Pezim had not even been drilling for gas; he had been drilling for sulphur. But gas is what he hit, and the shares of Stampede went from fifty cents to $27.

In this room, what they're talking about, finally, is money. It's the main event, even though the boss protests that he only uses it to keep score.

In front of The Pez there is a pack of king size filter cigarettes, with a Pez lighter sitting on it. Beside that is a deep-blue coffee mug, beautifully glazed, with Pezzaz scrawled across it in gold script. This is the name of one of The Pez's boisterous family of companies, currently a very active member of the family. Pezzaz is the reason that Joey Bishop, the famous comedian, is up here with his pal Murray Pezim right now, sitting behind Murray to the left, wearing an orange sweater. The sweater is so orange that if you were to bury it six feet underground, someone with poor eyesight would still be able to see it on a dull day. Joey is not a particularly retiring man, and he is telling everybody in the room how he met his pal Murray Pezim.

"It was eighteen years ago, and I was booked into a club here in Vancouver. Murray bought out the whole place, he bought out four hundred seats. Then he invited all his friends. He gave everybody a newspaper, and he told them, when Bishop comes on, open up the newspapers and start reading.

"So I come walking out, and all I see is four hundred people reaching down, picking up newspapers, opening them up and looking like they're all of a sudden more interested in the news than in the show. So I said: 'Just keep reading, everybody, and when you get to the Want Ads, let me know. 'Cause tomorrow I'll be looking for another job.' "

Pezzaz Productions Inc. is a wholly owned subsidiary of Pezamerica Resources Corporation. That is, it was until sometime in 1984. Only Murray Pezim could own a company called Pezamerica Resources. You simply cannot imagine a man like Conrad Black calling his imperial umbrella Blamerica Resources. Conrad Black's companies must be called names like Argus, Hollinger and Dominion. If you were to suggest to Conrad Black that he call one of his companies oh, let's say Blackeroo Inc., the suggestion would be received so coolly that you would be able to hear the ice cubes tumbling down the steps of No. 10 Toronto Street all the way to Vancouver. But out here, hell, Pezzaz sounds just great.

What Joey and Murray are up to is this.

Joey has gone around to all his famous friends — people like Sally Struthers, Buddy Hackett, Jonathan Winters — and paid them $10,000 each of Murray's money to record three-minute greetings on cassette. Some of these are funny birthday greetings, others are poignant get-well messages. All are from stars. They are packaged with splashy graphics and hung on rotating racks in drug stores and similar merchandising locations. The name: Greetings from the Stars.

You think this is a crazy idea? Greeting cards are an $8-billion-a-year business in North America. For one brief shining moment, Murray Pezim had a piece of that action with Greetings from the Stars.

You still think it's a crazy idea? Just listen to Shelly. Shelly is the slit-eyed New Yorker who's been yammering away from the sofa since someone let him in here at eight o'clock. You let him catch your eye for one-tenth of a second, and you're in for the Bloomingdale's story, like it or not. Saul has made that fatal blunder, and Saul's ear is now being chewed off by the merciless narrative of the all-conquering Shelly:

"To make a long story short, I finally got hold of the right guy at Bloomy's. I says, you gotta see this product. He says, I don't see the product. So I says, hold on, pal, you hold on. And I make a *package*. I'm talking *package,* Saul. I wrapped a bunch of the suckers up in bright red paper, and I tied a bunch of helium balloons to the package. Then I just carried in into his office."

Saul is so impressed with this story that he blinks. Just once, but then, this is Saul.

"I *fought,*" Shelly continues modestly. "You have to *fight*. I felt *so* good."

The TV reporter is not quite through. He has decided that there is no way he is leaving this place without a clip of Joey Bishop. His stock will go up so high when he gets back to the newsroom that he might even be able to sell it.

"All I'll say is, if once in your life," Joey Bishop is saying, "once, just once, you make a friend like Murray Pezim, then you'll be a lucky man. 'Cause what he does is remember what it was like *before* he made it."

In fact, this is true. The Pez has a weakness for charity. While all the rest of this activity is simmering around him, The Pez is dealing out instructions to a short, rug-topped man of indefatigable energy named Sammy. Sammy is looking after the seating and other arrangements for an extravaganza which is to be called, ahem, The Pezzaz Festival: "October the thirteenth, write it down," Murray orders with a grin.

The purpose of The Pezzaz Festival is to raise money for underprivileged kids, and as The Pez will tell anyone who will listen, even if he doesn't have a TV camera rolling at his side, "Listen, buddy, kids are our greatest natural resource, and I'm a guy that knows all about natural resources."

For Pezzaz, The Pez is bringing in Milton Berle, among others. You can bet that this is not the usual clutch of entertainers for Vancouver. Vancouver is a great place, but it ain't Las Vegas.

The way The Pez has conceived this night of a thousand stars is as a roast for his pal Peter Brown, the stock market whiz and jillionaire. Brown is young and handsome and dug so deeply into the fertile loam of the Vancouver aristocracy that he can take what's coming. What's coming will be knock-'em-dead lines like: "Peter who?" But he will probably laugh. Everyone will laugh. Because the line will be delivered by Milton Berle and will be costing them so much money to hear. While we are all sitting around trying not to hear the looped track of Shelly's play-by-play for the fortieth time, Sammy trots in to report that he's sold another table. Five grand. Seats ten.

The real business of the office never stops while the other activities are awhirl. Murray nods at the news that Sammy has unloaded for $5,000 ten chairs around a table where people will sit down to consume ten bucks' worth of food. Probably he is glad that things are going so well, but he just nods, pops another fat blueberry into his mouth and says, "Let's have a look at that bond market."

When the day begins at six-thirty, there is before Murray on his desk a large bowl filled with these enormous blueberries and several peaches. Now, The Pez does not just hump his chair up to the desk and start to shovel this stuff away. His concentration is required by both telephones, a dozen people, half a dozen monitors and a towering pile of correspondence and documents needing his signature. He must somehow get through all this while the rest of the dance is pounding across his stage. The fruit just disappears berry by berry, peach by peach, a nibble here, a bite there, chomp, work, chomp, work.

Later, an hour into the day, another plate of fruit is brought in by Susan. This is more elaborate. There are apples and other pedestrian members of the Canadian fruit family, but there are also papayas, pineapples, mangos and a tall, tumbling pile of big, black, round, glistening olives. Of course, this is not all that The Pez eats. The Pez is not some kind of vegetarian hippy. The Pez is not a fruit fly. You do not sit behind a desk as cheques for a million bucks tear in and out of your bank like speed freaks and make it through the day on a bowl of goddamn *blue*berries. A man has to eat if a man wants to deal that kind of bread every single hour of his livelong day and not go all wonky. So about ten o'clock, The Pez orders some real food.

"Hey! Where's Stan? Get Stan in here, okay, buddy?"

Stan does not so much rush into a room as manifest himself. Suddenly Stan is *there,* having entered with no more fanfare than a stalking cat. This is not because Stan is a slight person. If Stan wished, he could enter the room simply by walking through one of the walls. Stan looks as if he might have been made of Swedish steel and just covered with skin afterwards so that people in the street would not run away from him. People do not run away from Stan, but they edge slightly to the left or the right when they approach him along a corridor.

Stan is by way of being a bit of a bodyguard.

He is about six-foot-three, and maybe two hundred pounds. He is wearing white trousers that are very tight and show off the shape of his legs. They are well shaped, packed with muscle. Stan also wears a white T-shirt. This garment serves precisely the same function as the trousers: it displays to advantage those of Stan's attributes that are marketable. It also displays, on the left breast, the name and dates of the Pez Open, with the injunction: The Pez Says Be There. Stan never smiles.

Stan is by way of being a bit nervous-making.

"Hey, Stan," Murray calls. "Listen, make me a toasted bacon and tomato sandwich, okay, buddy? And listen, Stan, make the bacon crisp, okay, buddy? And Stan, put a fried egg in there too, right?"

Stan deals out a few coffees around the desk before beaming himself out of the room to perform the cook functions attached to his position with the Pezim organization. As he hands the big, blue, gold-scrawled mugs around there is plenty of opportunity to witness the rippling forearm and the bursting bicep as these are thrust past your nose to reach someone else. Stan presses weights. Stan was Mr. Poland in 1975.

If you are a visitor here, The Pez tells you to go ahead and ask for anything you want. If you are hungry, you do. You tell Stan that you want the same as Murray, please, if that would be okay. Stan looks at you as if, sure, that would be okay, as long as he can deliver it to you wrapped in a fist. But he does not say this, he just eases himself out of the room, not having to step *around* anyone, and disappears. Murray's sandwich arrives quickly. The visitor's sandwich does not arrive at all. No one notices this but the visitor. And Stan.

Later in the morning, if you decide to go hunting for the washroom, you tiptoe out of action central. This is unnecessary. You could jump out of Pezim's office on a pogo stick made of I-beams, and they would think you were trying to sell a new product. The washroom, you are told, is down the hall, first door on the right, the door just before the kitchen.

You almost walk past the door to the washroom, because it is just a panel of wood among other panels of wood. The handle is very discreet. This is not *every*one's can in here, you understand. You would still be looking for the can were it not for the fact that you notice the kitchen. There is a very well dressed man in the kitchen, standing before the microwave oven. He is swearing softly but with great fluency. There is evidently some disagreement with the microwave which this gentleman feels has gone beyond reasonable interface, and he is unburdening his soul. It is this little kitchen scene that alerts you to the fact that you have passed the washroom door.

Nothing you have been through in your middling span of years will have prepared you for what your eyes behold when you push through that panel of wood and step into the awesome world of The Pez's can.

It is a small world that you enter, no more than eight feet square. But the walls have been sheeted with mirrors. Everywhere you look, your own astonished face stares back at you, wondering how on earth a dash for the privy has landed you in the midst of such perfect narcissism. The floors are broadloomed wall-to-wall in white wool with a pile so deep that you would probably be wise to cross it on snowshoes. There is a little sink set diagonally into a corner, and a counter running along one wall is loaded with white towels. But all of this is no more remarkable than a country outhouse compared with what you see laid out before you in the centre of the room.

Shelly.

For once in the few hours since you have met Shelly, he is not trying to retail his account of the perilous and bold-hearted mission into Bloomingdale's. All that Shelly is saying, laid out here, is "Aaaarrrgggh!"

Shelly is stripped to the waist and lying on his stomach on a high table. He is still wearing the trousers of his shiny grey suit and still wearing shoes. This makes him look a little incongruous. Surely, you think, he ought to have taken his shoes off.

Shelly's back is covered with an oily shine. It smells the way health clubs smell, with that astringent gymnasium scent that suggests you are probably unfit. Standing over Shelly, his face set in happy concentration, is Stan. Stan prods around a bit with his fingers and then, apparently finding what he seeks, digs in for a handful.

"Aaaaarrrrgggh!"

If this is what Stan does with his hands between coffee runs and turns at the toaster, you are glad he did not fix your sandwich. The idea of wolfing down something that has been connected in any way with the victor of Bloomingdale's is more revolting than discreet expression can encompass.

To get to the actual toilet section of this Roman facility you must circumnavigate the groaning Shelly at the shoe end and make for a section of mirror that is slightly ajar on the far wall. It is a good thing that it is open a crack, for otherwise you would have had to ask Stan where on earth the door was. And it would be a pity to interrupt Stan when he is in the middle of one of the few tasks which obviously affords him some enjoyment.

Back in the office, Joey Bishop is starting to get antsy. He has an interview laid on with Merv Griffin later that afternoon in Burbank. The interview is to promote their new Greetings from the Stars. Joey has already had spots on NBC's "Today Show," and on one of those New York stations with more viewers than most networks. But Merv is not to be passed over, and Joey is wondering how they are all going to get down there on time if The Pez proposes to stay glued to his monitor busying himself with such trifles as a sharp rally for gold in London.

The Pez laughs. "Joey thinks this is just watching TV."

But although he has time to address a little drollery to his friend, The Pez does not for one second take his eyes off that screen while the gold rally is underway. Gold actually jumps $2 an ounce in a few hours, and downstairs, on the floor of the exchange that gold built and gold fuels, some people are starting to make rash decisions about their futures. Some people are starting to think in terms of a massive rally and some people are starting to think in terms of $500 an ounce. Some people are foolish, for gold on this summer day in 1984 is around $375, and if anybody could sniff a pole vault to $500, it would be the guy upstairs on the fifteenth floor with the friend at his shoulder nagging him about Burbank.

"We'll just watch this for a couple minutes, okay, Joey? Hey, Saul, look at gold. Let's just watch it for a couple of minutes."

This is telling the snake to pay attention to the mouse. Saul, don't worry, has been watching gold since its first burp. Saul is watching gold and Larry is watching gold and everyone in that whole building with two eyes and a monitor, which is everyone, is watching gold. Dan Rather could be standing downstairs in the lobby ready to go live into the CBS News about Greetings from the Stars, he could damn well wait till gold had made up its mind.

This is part of the fascination of a major market that is so frankly speculative, especially when that market is tied to the price of something as exotic and mysterious as gold. What happens when gold moves a point or two up is that some smart fellow, sitting in his lavish Vancouver office, starts to think that maybe this is the time to move a little of that dried-up, worthless, wrung-out gold mine he bought for a couple of grand. He calls his associate on the street, and he says, "Let's trade Fond Hope Mines, 'kay?"

And his brother in arms says, " 'Kay."

What often happens is this:

The friend places an order with his broker to buy Fond Hope, say 200,000 shares. The broker knows what's going on, but bites his lip and buys. Fond Hope jumps an inch on the board. Another friend places the same order with *his* broker. Fond Hope jumps another couple of points. The original owner buys back, and Fond Hope jumps again. This is called wash trading, where the buyer and the seller are the same person, or *de facto* the same person, and wash trading is illegal. It is illegal because it gives people the wrong idea. It gives people the idea that there is something *to* Fond Hope, because look at all the people who seem to want to own it. Eventually someone other than the original owner and his friends jumps onto the rising elevator of Fond Hope Mines. That is when the original owner and his friends get off. Fond Hope, with nothing on the board but an offer to sell, all of a sudden takes a ride down to the first floor again, with no one but the *new* owner aboard for the unhappy descent.

This does not happen to people like Murray Pezim. Maybe once, when Murray was learning that first lesson in the stock market back in 1949, but not now. Now Murray and his associates just watch for a while to see if gold is going to stick or slide.

"Don't worry, Joey. We'll get to L. A. in time."

To Joey, Murray is just watching an appallingly boring TV show featuring a string of changing numbers. But let us say you have an important position in a company that owns half of an ore deposit sitting in the bush alongside the Trans-Canada Highway 125 miles east of Thunder Bay, Ontario, at a place called Hemlo. And let us say there are 10 million ounces of gold in that rock of yours. Your half is 5 million ounces. If gold goes up $2 an ounce, then $10 million has just plopped into your pocket. If gold drops $2 an ounce, then you are, at a stroke, a man with $10 million less than he had yesterday.

This is some TV show, Joey.

Today gold settles. It drifts back to the same price it started the day with. Someone in London or Zurich has been playing games, having a little fun, picking up two or three hundred million on a flutter. It may have been the Russians. It may have been the South Africans. Whoever it was will have had their reasons for pushing the price up $2 for a couple of hours. A little nervousness about Namibia, maybe. Maybe some French ship with a lot more uranium than it should ever have had aboard has gone down off England, and the NATO allies have found out, and the Russians are not going to get that uranium, and the French are not going to get paid for it. The franc plummets, some Paris bankers scramble for gold, and the French take an afternoon bath. Too bad. They shouldn't have been selling the Russians that uranium anyway.

Another thing. You look for whoever came out of the afternoon loaded, you may find someone who knew about a certain shipload of uranium. And who sank it. Just a thought.

Seven

The Pez Aloft

Papers are being shuffled on the fifteenth floor now and stuffed into attaché cases. The Pez's own case is a mighty affair of shiny black patent leather, stamped with an alligator-hide pattern and bearing a brass plaque with PEZ engraved on it. The Pez is taking Joey Bishop to Burbank and is getting ready to do a little business while he's down there. Marshalled at the ready to assist the leader in his departure is the sturdy retainer, Shirley.

"Shirley, what have I got on my Gold Card now?"

"You're clear on the Gold Card. You can go to $50,000 on it."

" 'Kay. Get me $2,000 American cash, okay?"

"Sure. $2,000 American." Shirley is back with the two grand in minutes and The Pez just dumps it into the briefcase. Petty cash.

"Shirley, everything booked at the Beverley Hilton?"

"Everything's booked. Two suites. Two bedrooms apiece. They're waiting."

"Good. Limo?"

"Limo's waiting."

"Good. Thanks, Shirley."

Shirley staggers out of the room under a few tons of signed correspondence while Susan organizes another pile for The Pez to scrawl through in a final blur of activity.

"I hope all this stuff's in order."

"It's all in order."

''Good.'' But The Pez reads everything anyway. Naturally he reads it fast, but nevertheless he reads it. You can see his eyes tear over the pages as, one by one, he reads them through, signs them, and they are whipped off the pile so he can read the next one.

''Don't forget to load that little stand, the one that holds the cassette upright so the camera can see it while Joey's talking to Merv.'' The Pez may be reading and The Pez may have just watched a gain of $10 million sink back into thin air, but The Pez can still think ahead. ''And throw in some of those golf shirts, a few sweaters and a bunch of those pens, the new ones.''

These last items are really quite beautiful. They are very pricey pens indeed, a long way from the usual plastic throwaways. These are made of satiny black metal, and they are stamped in gold with a heart, followed by ''The Pez.'' The word *Pez* is also encased in a clear plastic bubble atop the pen. The nib, where the ball point emerges, is gold, as is the clasp that holds the pen to your pocket.

Saul says: ''Watch it with those new ones, Murray. We only got five hundred and you've been handing them out like peanuts. You're supposed to keep them for Pezzaz.''

''Yeah, okay,'' says Murray.

His eyes stray to the box sitting in front of the visitor. This handsome gold box holds one of the new pens. The visitor slips it into his attaché case. Fast. The visitor wants to keep it.

A short, dark man named Carlos appears with a porter's two-wheeled baggage carrier and loads it with boxes of tapes, display stands, sweaters, shirts and Murray's attaché case. He wheels it towards the door, and half the office trails off in his wake. Murray says to his visitor:

''Look. Why don't you just come on down to L. A. The plane could turn around and bring you right back. Only take a couple hours. That's probably the simplest idea.'' The simplicity of this idea appeals to everyone. Even Saul seems to think it's a good idea. Certainly the visitor does. When you are used to doing your travelling on the Toronto subway, the idea of blasting off to L. A. with Murray Pezim and Joey Bishop is an exhilarating prospect.

It is a crowded elevator that descends to the parking garage in the basement of the exchange tower, stopping all the way. Everyone in the building knows Pezim, naturally. And naturally, Murray pretends he knows everybody else.

''Hi, Murray.''

''Hi, buddy. How are ya?''

''Fine, how's it going?''

"Fine."
"Great. See ya, Murray."
"Hi, Murray."
"Yeah, see ya. Hey, hi."
"How are ya, Murray?"
"Fine, just fine. You?"
"Great. See you, Murray."
"Hey. Hi, Murray."
"Okay, buddy. See you later. Hi, buddy, how are ya?"

And so on, down sixteen floors to the basement parking garage. The long black Cadillac is right there, in the parking space closest to the elevator. Carlos begins to pile the luggage into the trunk, and the rest of the cargo — people — sort themselves out for the ride to the airport. There is Murray, Murray's guest the visitor, Joey Bishop, Shelly and Alex. Alex is Murray's stockbroker, and he climbed aboard somewhere on the elevator ride down. How he knew which elevator contained Murray, one does not know. But it could be that he simply stuck his head against the various closed doors until he heard "Hi, buddy" drifting down the shaft towards him.

Everyone's role here is more or less clear, except Shelly's. Shelly has fixed himself to Murray and Joey as if he were fastened there with a steel weld. Sure he sold Bloomy's, but by the time the Cadillac is purring through Vancouver on its way to the airport, Bloomingdale's has sunk so far back into the pages of history that it is being taught at the Harvard Business School. For by *now,* someone has sold Macy's. And they are beginning to close on Gimbel's. Bloomingdale's is okay, but Shelly's position as the undisputed wholesaling king of Greetings from the Stars has not survived the morning, and now it looks as if Shelly just wants to hold on to his friends at the top any way he can. If that means a flip to L. A., well lemme on.

The black Cadillac flashes through the gates and onto the broad stretch of tarmac at Vancouver airport before the man at the gate can even step from the guardhouse. The Pez is in a hurry, son. This corner of the airport — a mile across the field from the glass-fronted passenger terminal where lesser folk arrive and depart on a timetable set by someone in an airline office — is full of private aircraft. They shuttle in and out of here all day on the business of the great corporations based in the golden province.

MacMillan Bloedel has a whole section just to itself. MacBlo is one of the mightiest forestry corporations in the world, and its fleet of aircraft is the size of a small air force. There are at least a dozen light planes painted with the company colours. There are helicopters. There is a pair of huge, ungainly flying boats that look like crippled geese as they waddle and weave across the tarmac. These fascinating craft may resemble museum relics, but they are real Clydesdales, ploughing aloft into all kinds of weather to ferry the technicians and managers and men who keep the forests ringing to the sound of MacBlo's labours.

There is more than a little irony to our passage past the busy MacBlo hangars. MacBlo is the giant that almost killed a giant. In 1981, after a bitter and ruinously expensive takeover battle, MacBlo was seized by the Toronto-based natural resources empire of Noranda Incorporated. Almost as soon as Noranda had MacBlo under its wing, the whole resources industry plunged into a slough, and the downturning fortunes of MacBlo contributed heavily to the drag that pulled the great Noranda down. Along with other possessions, MacBlo was bleeding Noranda white. Frantic for some profitable enterprise, Noranda surveyed the field, and her eye came to rest on Hemlo. Shouldering aside the indifferent neighbours of corporate Toronto, Noranda grabbed for the Hemlo gold camp and took 50 per cent of the action on The Golden Giant deposit.

The guy with the beat-up honker who is gliding past all those MacBlo planes, why, he's the guy who *made* Hemlo. The lesson is, never judge a man by the number of planes he owns. They may be costing him so much dough that next time you see him he'll be waiting for the bus.

And there's another irony.

In the fall of 1983 Murray Pezim made a deal with Noranda. Alf Powis, Noranda's chairman, announced to a dumbfounded mining and financial community that Pezim-controlled companies would be financing explorations pinpointed by Noranda. The mining promoter from Vancouver, the butcher's kid from the Jewish ghetto west of Spadina Avenue, a man whose name, if uttered in the precincts of the Toronto Club would have brought either laughter or icy stares, this man was going to *bankroll Noranda,* for God's sake!

This is how the deal works:

Noranda's subsidiary, Noranda Exploration Limited (Norex), will provide the geological and technical expertise to locate promising new properties, and Pezim's junior companies will then provide the financing to pay for the drilling program to establish whether the properties are worth a major commitment.

"It's like the original grubstake business," says John Harvey, president of Norex. "The person with the money finances the person with the expertise. In the past, Noranda has had to rely on exploration funding coming out of income, debt-financing or whatever, but because of difficult times in recent years there has not been a lot of money available for exploration. The deal with Pezim will give Noranda more exposure as far as new properties are concerned."

There's another way to express this: The Pez had the money for holes; Noranda didn't have the money for holes.

What the deal gave Murray Pezim was immediate respectability on the street. Pezim could now call himself a financier instead of a stock promoter. In fact, he does. Noranda expected Pezim to commit as much as $40 million to finance new drilling when they made the deal with him. According to the terms, if the partners struck pay dirt, then Noranda would arrange the senior financing necessary to get the mine into production. Senior financing, when you're drilling into the granite of the Pre-Cambrian Shield, means anywhere from $100 million to $250 million. The Pez feels the deal benefits the shareholders of both partners.

"*My* people are voluntary risk-takers. The people who buy shares in Noranda do so for dividends, not speculative opportunities. But if Noranda is selecting the properties and doing the work for a junior company, the speculator is going to have a lot better chance than with some company that spends its money dabbling on the surface of any old property. That stuff's all bull."

If all of this sounds too wonderful for words, then think again. Not very much in the world of business is particularly sweet. People are there to make money, and it is not often that they are thinking of someone else's prosperity while they go about it. Thus, the perfect symbiosis of the Pezim-Noranda deal strikes some analysts as bearing closer scrutiny. Sure, the reasoning goes, Noranda has had some rough weather. But was it so bad they need to go to Murray Pezim for small change? Here is the chilling observation made by one analyst, Don Scott of Bell Gouinlock Limited, in an interview with *The Toronto Star* when the deal was struck:

"It doesn't make any sense unless Noranda is trying to tie up Pezim."

Look across the back seat as the limousine sways over the last few hundred feet to the anonymous white hangar set off by itself in a corner of the field. The Pez has his wrist looped through the handstrap as he lounges back into the deep upholstery, chatting unconcernedly to Joey Bishop. Joey is griping at him for leaving their departure to the last minute, and The Pez is laughing, telling him to relax. It is the repartee of old friends, not worried about being misunderstood, not worried that a careless word will wound, certain of their ground and comfortable together. You wonder whether some quiet people who went to the same private school are not, right now, closeted in Toronto preparing to snap the rug out from under this big, happy man. They have plenty of reasons to hate him, but two would be enough.

One: he is not one of them.

Two: he took a gold mine out from before their eyes.

Wealth can look seamless. In reality it is often riddled with a thousand tiny cracks, and only the man trying to hold on to the money knows for certain where they all are. The Pez knowns this better than most men. He knows it from his own past.

Pezim's father was an immigrant, a Romanian-born Jew who settled in Toronto with not much more than a headful of hope and the determination of a locomotive. He managed to establish two drug stores, and when Prohibition hit in the 1920s, he made himself a fortune selling medicinal alcohol to people who teetered into the store with their tongues trailing behind them. Pezim Sr. did so well that he installed his family in one of the big, boxy, three-storey mansions that crowd along Palmerston Avenue between Bloor and Harbord. From the outside it must have looked so safe. But all it took was the stock market crash of 1929 and the ensuing depression to put the young Murray Pezim into a butcher shop in the ghetto with a lot of hundred-hour weeks staring him in the face.

Cut to 1981. The kid from the butcher's shop has just plunked down $630,000 for a lavish condo on Vancouver's English Bay. That's $630,000 *cash*. But then the stock market collapsed, and all of a sudden Murray Pezim is worth $20 million less than he was the month before. And to make things worse, there are some very slit-eyed men at the door waving a wad of papers that say the guy in the condo has committed stock fraud. If guilty, he's looking at jail.

But The Pez is acquitted, and soon he's back at work, building up his fortune again. By 1983 estimates put his net worth at $40 million. And so to this day, riding in his limo, Murray Pezim is looking as secure as a brick of solid gold. But he knows the truth. He knows that nothing is safe.

The car pulls round in a circle and eases to a stop beside a trim aircraft. The engines are already running up on the Hawker-Siddley twin-jet, and the gangway stairs at the forward end of the cabin are folded down and waiting. The pilot, Jim Barrett, is an air force veteran who has flown everything from fighters to the fat, lumbering Cosmopolitans that the armed forces still maintain in their VIP squadron. For the last few years of his air force career, Barrett carted planeloads of generals back and forth between Canada and its European bases; or he carted generals between Canada and our peacekeeping stations in Cyprus and elsewhere. Generals are not much fun. Barrett likes flying The Pez better.

The copilot, Norman Harriman, was a first officer on jets for CP Air, flying the South American routes which follow the west coast and fetch up in places like Lima. You have to be very easily entertained to find civilian airline flying interesting. You always go to the same places. You always go at the same times. You always go at the same altitudes. You always go at the same speed. And you always carry around the same loads of people, complaining about the food.

"With Mr. Pezim, you never know where you're going," Barrett is saying. "We fly to Mexico, Arizona, all over the States and Canada, never the same place twice in a row. To Mr. Pezim, this is just taking the car out for a spin. Down to L. A., hell, it's only two and a half hours."

The Hawker-Siddley can seat seven passengers in its snug little cabin, and it can seat them in considerable luxury. Appropriately, the colour scheme is gold. The seats are covered in a thick, rich, gold material, and the bulkheads are decorated with coppery panels. Forward of the passengers' cabin, between the passengers and the cockpit, there is a little cloakroom. At the very rear of the cabin is the washroom, and just ahead of the washroom, the galley.

The galley is small but, on this flight, it is stuffed with enough food to feed a Cosmo-full of generals. For The Pez and his tight little entourage, it is a banquet.

There are sandwiches crammed with thin slices of black forest ham and slabs of Swiss cheese. Fresh little ovals of rye bread are skewered with toothpicks and fastened onto either end of little piles of rare roast beef. To call these sandwiches is to call a semi-trailer a truck. There are the ever-present trays loaded with fruit: papaya, mango, watermelon, canteloupe, grapefruit. There are platters heaped with Cheddar, Lancashire, Brie, Camembert, Stilton. There are little pots of mustard and mayonnaise for the sandwiches. There are olives and pickles and celery and carrots and thin, pale tips of asparagus. There are coffee and tea and soft drinks and a cooler clinking with bottles of chilling wine.

The Hawker-Siddley scoots along the concrete and darts aloft with no more effort than a swallow in the summer air. Alex the stockbroker and Murray's guest are sitting in the two front seats. There is one seat on each side of the aisle, for three rows. Then there is the galley on the starboard side and a long bench on the port side. Joey Bishop and Shelly are sitting in the third row, the aisle between them, and The Pez is lounging along the bench, his briefcase open beside him, leafing through some papers. Alex is a veteran of these flights, which is how he knew enough to sit in the very first row. The guest just lucked out. It works like this:

Murray always takes the bench, because the bench has most room. This leaves the front three rows for the other passengers. Now, the first-row seats face aft, staring straight into the second-row seats. Thus, if you pile into one of the front seats, with your back to the front of the aircraft, anyone taking a second-row seat must sit face-to-face, knee-to-knee with you. If there is enough room, few people will do this, as Alex knows. People would rather go to the third row. This leaves the lucky grabber of the first row with, in effect, two rows to himself. In a plane this size, that is a nice enough edge, as Alex knows. Alex fills you in on all this strategy of executive flying, for he is an open, friendly young man. Just how it is that Alex comes to be Murray's broker, however, is another story, and Alex is not prepared to share it with you quite as selflessly as he will the lesson in seat tactics. He will confide that he is just one of Murray's brokers, not the only one, and that is it.

Alex wears gold chains, gold rings, a gold watch and a sweater with no shirt underneath. His pants are draped very loosely right to the ground, in a sort of Italo-Californian marriage of natty comfort and plain sloppiness.

The jet climbs fast out of the airfield and Vancouver begins to drop beneath the starboard wing. In a few minutes we are at five thousand feet and still ascending rapidly. The Strait of Georgia gleams away northward between the mountains of the Coastal Range and the steeply rising hills of Vancouver Island. From Horseshoe Bay the ferry sails out into the strait, shaping its course for Nanaimo, visible from this altitude on the Island's east coast. Slowly, still climbing, the jet banks away to the left, heading onto the course south that will carry us down through three states to the dozen sprawling square miles of pavement where we will set down: Los Angeles International Airport. Now Victoria is visible, and the ferries that snake their way from Tsawwassen on the mainland through the Gulf Islands to Swartz Bay near Victoria carve their sparkling trails of sunlight upon the water.

We climb steadily until we reach 37,000 feet, seven miles up, and head south at 450 miles an hour. First the gaping, ragged crater of Mount St. Helens passes beneath, then the shining peak of Mount Hood. The jet leaves the Northwest behind, picking up the cordillera of the Sierras. Towering thunderheads stand up along the High Sierra, some of them a hundred miles high. The pilots watch them closely on the radar, keeping as far as fifty miles out of their way because of their massive hail stones.

Back on the rear bench, The Pez has had a good nosh, and he's ready to talk about his gold mine.

"You know, you don't just sit around waiting for some old guy on a mule to come riding into your office and whack a nugget down," Pezim says, shuffling some of his papers together and stuffing them back into his attaché case. He snaps the case shut and stares out the window at the distant storm clouds, thinking about his gold mine.

"It was just the most beautiful engineer's report I'd ever seen. Steve Snelgrove brought me that property. He brought me in this beautiful engineering report on the geology out there, and I was very impressed. Very impressed.

"I closed the deal with Steve right there."

This is more or less what happened, but it is not *exactly* what happened. What happened was that Nell Dragovan closed the deal, but she could never have arranged the financing herself. Not then. She could now, easily, but not then. So Murray helped out.

That is how simple it is: Murray Pezim liked the report Steve Snelgrove brought to him in Vancouver, when Steve couldn't get a phone call returned in Toronto. Murray Pezim out on the coast thought he had a beautiful report there, and Murray ended up with the first piece of what could easily be the richest goldfield in the Americas.

For quite a while, it looked as if Murray Pezim was the only one who thought he had anything more beautiful than a nice report for his forty grand and his 300,000 shares of — what was it? — Corona Resources.

"At that point we knew we had a good prospect. That's all. But at least we knew for sure we had a good prospect. It was 80,000 tons of very marginal ore, grading say maybe 20,000 ounces of gold in the whole thing. That's all we knew we had.

"I raised $1.2 million right away for drilling. We drilled seventy-five holes and ended up with 300,000 tons grading 0.20. Everybody said, 'Quit. You're never going to get anything in there.' "

What 300,000 tons of ore grading 0.20 means is that for every ton of ore there is one-fifth of an ounce of gold. What The Pez had, then, was 60,000 ounces of gold. Now, 60,000 ounces of gold, even if it is worth only $350 an ounce, is still $21 million worth of gold. Trouble is, the $21 million worth of gold is not lying around on the surface of the ground in bars. It is underneath some exceptionally tough rock, and to get it out you must dig a hole. And if that hole is going to cost you $100 million to dig, well sir, the gold stays right where it is.

"But then we drilled Hole 76."

Oh yes. Then they drilled Hole 76. And Hole 76 nicked the very edge, *just caught a tip* of that great, north-sloping, east-trending slab of gold-bearing rock.

Hemlo.

Gold rush.

Eight

Rush on The Hemlo

There was nothing but a lot of freezing, snow-blown bush to mark the site of the future gold camp when geologist Dave Bell arrived on The Hemlo in January of 1981. Cutting a road in from the highway, Bell and his crew set up their diamond drills and started to probe the frozen rock. If anyone in the powerful Toronto mining establishment even knew he was there, they showed little concern for the knowledge. Who was behind the geologist? A two-bit Vancouver-based resources company called Corona Resources and a nobody named Nell Dragovan. Even the name Murray Pezim would have stirred up little more than a sneer in early 1981.

Bell began his drilling program without any of the usual back-up that a modern geologist commands. There were no geophysical and no geochemical studies to help him pinpoint anomalies in the rock. Geologists love anomalies; anomalies mean minerals.

But Bell plunged into the bush anyway and began to drill the area of the claims where gold had previously shown up. Before long he had outlined 750,000 tons of gold-bearing rock. But the grade was too low. Gold prices were falling, and the cores were grading only around 0.10 ounces of gold per ton. Not rich enough to mine by a long shot. But Bell kept at it.

The zone that Bell was drilling later became known as the west zone of the Corona deposit. Bell stuck with the west zone for seventy holes, until it became obvious that he had outlined all there was at that location. He had probed as far down as he could with his drills, and he had moved along the strike, or gold-bearing zone, as far as he could.

At this point most geologists would have given up. Seventy holes is a lot of drilling when your results are grading consistently too low for the rock to qualify as ore. But Dave Bell had a hunch. He believed there was gold in the geology of The Hemlo. To find it, he had to look a little further, drill a few more holes. This is where the backers became very important.

Bell had warned Dragovan and Pezim that a proper assessment of the Hemlo claims would demand a steady and fearless hand on the chequebook. It must have taken an enormous amount of nerve, pouring hard cash into a growing collection of holes in the ground, but that is just what Pezim and Dragovan did. Hole after disappointing hole, they stuck by Bell and kept the money coming. This is what separates the hot-air school of promoting from the honest-to-God grubstakers who lay down hard cash.

For seventy holes they plugged away at the west zone. John Dadds, a geological technician, helped Bell on site. The two men shared the exacting job of examining and logging the core as it came out of the hole, and they conferred on spotting the holes.

And then Dave Bell took the gamble.

The eastern edge of the zone they were drilling was defined by a diabase dike, a thin wall of hard, dark-coloured igneous rock, plunging down into the surrounding rock. On Hole 71, Dave Bell took the drills beyond the dike for the first time. His hopes were very high. The men on the drills were excited. Maybe at last they would find something. Maybe they would find themselves on a gold camp after all.

Hole 71 came up blank.

The next four holes came up blank.

And then they drilled Hole 76.

Hole 76 would turn out to be one of the most important holes that any geologist had ever drilled in Canada. In the history of Canadian gold mining, Hole 76 will have a unique place. Hole 76 bit into the richest zone of ore that Canadian gold mining has ever known, intersecting that steeply angled shelf of fabulous rock at a depth of 336.5 feet. For a distance of 10.5 feet the drill bit through the ore and then out into the schist beyond. The 10.5 feet of ore graded 0.209 ounces of gold per ton.

Of course, there was no whooping and dancing around the drill rig when they pulled the core out of Hole 76. For no one knew what they had intersected. By the time Bell drilled Hole 76, they were well into the busy spring exploration season, and there was a long wait for the assay results. But Dragovan and Pezim stuck in behind Bell, and Bell stepped boldly out into the new zone, ranging four hundred feet farther east with each subsequent hole.

Bell cut into the discovery zone again on Hole 78, pulling *15 feet of ore grading 0.343 ounces of gold per ton*. On the next five holes Bell went far beyond the ore zone, driving his drills down into the rock in a daring search for the slumbering gold he knew was there but which he did not realize he had already found.

And then the assays for Hole 76 came back.

Bell was galvanized. He immediately reported the results to Corona in Vancouver, and Pezim and Dragovan authorized more drilling. They had grit, those two desk-bound wheeler-dealers, and finally they were beginning to see a little light break along the horizon. The diamond drills went back to Hole 76 and started to probe the ground around it. Almost right away the size of the deposit beneath the anxious crew began to expand. Hole 86 started the new probing that would ultimately define the extent of the ore body in the east zone of the Corona deposit, and with Hole 86 the assays began to come in high enough to startle investors out of their lethargy. Even in Toronto they began to watch Hemlo.

By the time Hole 120 went down, Dave Bell had outlined 250,000 tons of ore grading 0.25 ounces of gold per ton. Back in Vancouver, Murray Pezim and Nell Dragovan were starting to move their stock. It was not hard. All they had to do was make sure those press releases kept on coming, and they were both past masters at the art. Corona stock started to climb.

The fact that Dave Bell had outlined a quarter of a million tons of ore grading 0.25 was not by itself the kind of information to set the mining community ablaze. A deposit that size, at say $350 an ounce, is only about $22 million worth of gold, and that is not enough to open a mine in mean rock country like Hemlo. But what the discovery *did* mean was that Bell's theory was correct.

All along, Bell had maintained that the place where two different kinds of rock — volcanics and sediments — met would be a good place to look for distinct ore bodies. The place where two different kinds of rock meets like this is called a stratagraphic contact, and it was this contact that Bell had been following with his drills. The stratagraphic horizon for the whole Hemlo was immense, and Bell had only probed a small part of it when he outlined the rich east zone of the Corona property.

Dave Bell started to send down the deep probes. He knew the ore plunged off to the north and he wanted to see how far. The first deep hole on the east zone cut into the ore at a depth of 839 feet and came out the other side at 901 feet. This doesn't mean that the ore zone at that depth was 62 feet wide, for the drills would slant into the ore at an angle. Nevertheless, the results seemed to indicate that the ore body widened out at depth. By September, drilling along 700 feet of the strike length indicated a true width for the deposit of between 15 and 20 feet.

Meanwhile Corona was no longer alone at Hemlo. Lac Minerals had optioned the eleven Williams patented claims from the widow of the original claims-holder. (These are the claims that are at issue in an action between Corona and Lac before the Supreme Court of Ontario.) As early as April of 1981, when Bell's drilling for Corona was still in its early stages, Lac crews staked seven hundred claims on the Hemlo.

''We had the property in our minds long before that,'' Lac's vice-president of exploration, Dennis Sheehan, told *The Northern Miner*.

But Corona was doing the drilling, and Corona's stock was rising. By September Murray Pezim had organized a new infusion of funds for Corona. Pez Resources, one of Pezim's companies, committed itself to purchase up to a million shares of Corona Resources for a hefty $20 a share. It was not as hard to organize this deal as it might sound. The Pez was chairman of both companies.

On The Hemlo, Bell kept drilling.

There was now so much interest in Hemlo and so much activity in the Corona stock that a Vancouver brokerage firm commissioned an independent report on Corona. Corona had traded up to $32.50 a share, and Continental Carlisle Douglas, the broker, wanted to find out whether its clients were paying $32.50 for well-prepared moose pasture, or whether Corona was really pulling the cores it claimed. The engineering firm of David S. Robertson and Associates gave Corona top marks. All the results tabled were accurate, their report maintained. More than that, the report established reserve figures for the richer east zone, claiming it might contain as much as *803,845 tons of ore grading 0.31*. At late-1984 prices, that is close to $90 million worth of gold, and even on The Hemlo you can start a gold mine if there is that kind of money waiting inside it. The report concluded:

"The Hemlo deposit of Corona Resources is a remarkably continuous gold concentration which may be of paleo-placer origin. It has excellent tonnage potential and will likely exhibit a good consistency in grade. . . . the present study indicates the Hemlo deposit to be of strong economic potential."

The first major jumped.

Teck Corporation bought into Corona immediately, grabbing 200,000 treasury shares for $1 million cash. The money was earmarked, as part of an agreement between the two companies, for a detailed feasibility study of the Hemlo property. Subject to the results of the study, Teck would finance bringing a mine into production, earning thereby a 55 per cent interest in the property, with Corona retaining 45 per cent.

If you are wondering where the original prospectors were in all this trading and touting, don't worry. The 3 per cent royalty on net smelting was beginning, at that point, to look as if it might add up. And then there was the matter of a few hundred thousand shares of Corona stock. By the end of 1981, Corona was trading at $34 a share. On paper, Don McKinnon and John Larche were worth about $5 million apiece. And The Hemlo was just getting started.

Teck started to probe the Hemlo property right away. By June of 1982 Teck was reporting that the east zone contained 1.3 million tons of ore grading 0.30 ounces per ton. These figures, the mining company said, were "preliminary, open to extension, and probably conservative." Even at the low prices prevailing at the end of 1984, that was about $150 million worth of gold. The mineralized zone was eight hundred feet wide, striking in an east-west direction and diving down into the host rock from a depth of one hundred feet to a vertical depth of thirteen hundred feet. The ore zone was between six feet and twenty-five feet thick, with an average thickness of ten feet.

Then Lac began to drill Hemlo.

Peter Allen gets excited just recalling those early thrusts into the rock. Allen is the president of Lac Minerals, and a man who comes to gold mining from out of a legendary past. It was Allen's father, John, who flew down to the Bahamas when Sir Harry Oakes was murdered, and who bought Lakeshore Mines right on the spot from Oakes's widow.

History has repeated itself on the Hemlo goldfield. The heart of the Lac claims at Hemlo is the eleven-claim block known as the Williams Option, the property patented in 1945 by Dr. J. K. Williams of Maryland. When The Hemlo started to heat up, Lac went looking for the owner. And they found her. Mrs. Lola Williams, the doctor's widow, lives in Salisbury, Maryland, where, well past the age when most people retire, she still works as a registered nurse.

Lac opened negotiations with Mrs. Williams on 3 July 1981. Just over three weeks later, on 28 July, Lola Williams formally accepted an offer of $250,000 and a net smelting royalty of 1.5 per cent. Lac handed over its down payment of $25,000 and moved onto the Williams claims.

Whether Lac stays there or not is another matter.

In October of 1985 the Ontario Supreme Court will begin to hear an action brought by Corona against Lac, in which Corona claims that it, and not Lac, rightfully owns the Williams claims. Corona maintains that Lac used privileged information to obtain the patented block. The suit is the largest in terms of value in the history of Canadian law. What is at stake is property worth billions, and the issue will be hotly contested when it finally comes to trial. Two of the mightiest law factories in the nation are squaring off to fight it out in court: McCarthy and McCarthy representing Corona, and Fraser and Beatty for Lac. If Peter Allen, Lac's chief executive, is a mortal man, then that lawsuit is the last thing he thinks about when he goes to bed at night, and the first thing he thinks about in the morning.

For sheer drama, Peter Allen wins the executive office sweepstakes hands down. Lac's offices are on the twenty-first floor of the north tower of Toronto's Royal Bank Plaza. It is an appropriate place for a gold miner to hang his hat, for the Royal's towers are plated with a special glass infused with real gold. Lac's receptionist sits not so much at a desk as behind an arrangement of gleaming polished granite slabs, reflecting the ceiling lights. On one wall a many-textured tapestry mural depicts a waterfall, and on another is painted a grey northern landscape. Beneath this landscape are scattered some of the minerals that Lac mines.

It is a short walk from the reception area to Allen's office. The carpets are a vivid green, marked in a spare design of gold. Access to Allen's office is through a pair of heavy doors whose handles are the polished horns of some exotic quadruped, curved and African-looking. They suggest that someone like a Ewing may dwell beyond the doors. They are wrong.

Allen is a tall, thin man with blond hair and very steady, very blue eyes. If central casting ever went looking for a WASP executive, they need look no farther than the twenty-first floor of the north tower of the Royal Bank Plaza. Allen is beautifully dressed in a quiet grey suit of a plaid so muted it can scarcely be heard above the air conditioning. His pale blue shirt is frayed just enough to let you know that in these parts, chum, Peter Allen is certain enough of his position that he does not have to throw anything out simply because it has reached the point where it might offend Zena Cherry, *The Globe and Mail*'s monitor of the rich and powerful.

The green cursor of a video terminal winks from the corner of a cabinet to the right and behind Allen. Beneath the screen is a single word, Reuter, the name of the international London-based wire service, a company that moves everything from news of the latest assassination in the Middle East to the spot price for gold in Zurich. Reuter has bureaus in every important city in the world, and since Reuter does, so does Peter Allen. Allen is trying to put the Hemlo gold camp into perspective.

Allen's expression alternates between furious disapproval and a surprising, boyish smile, as if he cannot make up his mind whether a writer ought to be burnt at the stake or treated like some harmless, genial lunatic from another, rather lovable, world. But when he gets onto Hemlo, everything is enthusiastic explanation. Out comes the paper; out comes the pen. And beneath the sketching hand of Peter Allen, executive, The Hemlo jumps to life.

In the spring of 1982 Lac conducted geophysical surveys on its property. These indicated a substantial deposit. The indications were correct. On the very first hole, Lac struck 80 feet of ore grading 0.18. That width indicated the best concentrations of ore on The Hemlo. Down into the rock Lac geologists sent their drills again and again and again. The results were astonishing. In only about twenty holes, Lac outlined an ore deposit of *almost two million tons grading 0.175*.

By the end of 1984 drilling on Lac's Williams property had outlined 52 million tons of ore containing 10.3 million ounces of gold. At 1984 prices, that was worth $5 billion.

Lac's plans call for North America's largest gold mine on the site, one that would produce about 450,000 ounces of gold a year. To do that, the operation would have to mine 6600 tons of ore a day. The target date to reach that level of mining is 1988–89. Lac will open with a 2200-ton-per-day operation in 1986.

Until he started to explain Lac's tremendous good fortune on The Hemlo, Allen's huge expanse of modernistic desk was broken only by a few neatly placed files. Now it is littered with pieces of paper thick with Allen's vivid illustrations and marked in his crabbed script.

"The significant thing about The Hemlo is that it shows that there is ore in these synergistic deposits."

Call them whatever you like; Dave Bell was right. In this weird clash of rock that surfaces at Hemlo, where Dave Bell believed there might be ore — rock worth mining — there was ore. And Lac had found it, too. And men poured into the bush.

There had already been a rush into the bush when news of Dave Bell's discovery of the Corona east zone leaked into the outside world. The stakers had moved in, taking the ground to the north and west of Bell's drills. They were following a geological formation known as the Heron Bay greenstone belt. To the east of Corona, Lac had staked its way out along the strike for miles, and Dennis Sheehan was telling reporters: "I think there will be more discoveries made in the camp. I don't think these [Lac's impressive tonnage on the first twenty holes] are isolated. We're quite hopeful about it." Lac controlled about 32,000 acres of Hemlo ground.

The rush to Hemlo became an avalanche. Prospectors from every bush camp and mining town in the country fell upon The Hemlo like an army of berserk piledrivers. Wood started to go into the ground so fast you'd have thought they were building houses in there. Professionals, amateurs, claim jumpers, bush thieves, crazies, semi-crazies, city boys, country boys, young boys and old boys, they hit the bush like the Allies at the Normandy beaches. There were a lot of pretty strange practices going down on The Hemlo when the staking rush heated up, and if you didn't have your claim walled off like East Berlin, well by God you'd want to keep a sharp eye on it.

Want a lesson in claim staking? Don McKinnon charged into the bush with a crew of a hundred men, and in the next twelve months he and his veteran wood-pounders had their tags nailed onto enough posts to encircle twelve thousand claims. That is a syndicate controlling 800,000 acres of ground. Most of the ground taken by Don and his men was within forty miles of the Corona strike. All told, there were something like two million claims staked in the area around Hemlo in the wild days of the staking rush. Of course, there were plenty of claims staked elsewhere, too. Prospectors were calling anything within two hundred miles of the strike "Hemlo" ground, and if you didn't watch what you were buying, well then you might just find yourself with a piece of land so far from Nell and Murray's patch you'd be able to get to it on a Toronto city bus.

Meanwhile, Goliath Gold Mines and Golden Sceptre Resources, two members of the group of companies owned by Vancouver promoters Dick Hughes and Frank Lang, had exercised their options to pick up 156 claims to the north and west of the Corona ground. Goliath–Golden Sceptre hired Dave Bell, and Bell moved over onto the new ground. First results were unpromising and one drill, reaching down to test for an extension of the Lac deposit, hit a dike and had to be stopped. But it wasn't long until pay dirt.

Bell pulled a core with 98.2 feet of ore grading 0.256 ounces of gold per ton. This was the best hole drilled so far in the young gold camp, and after only five more holes, Goliath–Golden Sceptre announced a deposit of *2.5 million tons grading 0.25*. And then the heaviest artillery to date thundered its powerful message above the teeming Hemlo.

Noranda came aboard.

When Hemlo had begun to rise to fever pitch, Noranda had not been entirely asleep. It had been very quietly staking throughout the area. But when the Goliath–Golden Sceptre results came in, the mightiest natural resources company in the land felt it was time to take a position. When Noranda Inc. takes a position, that position is usually the front seat.

Noranda came in for 50 per cent of The Golden Giant, the fantastically rich deposit that Bell had outlined with five holes. The deal was that Noranda bring the property to production within two years. That two years was up in November of 1984, and part of the reason for the wild fury on The Hemlo has been that Noranda pays a penalty of $100,000 a month for every month the mine is overdue. But this is small change to Noranda. What Noranda wants is the gold.

Nine

Staking: The Rules

How does a prospector stake and hold a claim? In Canada the activities of prospectors are governed by provincial statute, and in Ontario the statute is the Mining Act. The Mining Act comes bound in a pale blue cover, and is 132 pages long. If you are planning to become a prospector, it is a good idea to learn what it says. Otherwise you may spend a lot of time traipsing around and hammering wood into the ground with no benefit other than that gained by a lot of strenuous exercise. If your aim is to keep fit, okay. If your aim is to get rich, read the Act.

The first thing a prospector must do is arm himself with maps. This is not so crucial in the developed and well-surveyed southern part of the province, where a claim must be described according to lot and concession number anyway. But in the north, few of the townships are surveyed into concessions or lots. The only lines that have been drawn onto the maps are the boundaries of the townships themselves. Even the use of the term *township,* implying as it does some human component, is misleading, for these are the ''unorganized'' townships, townships with no structure of government at all. In the north, the word *township* is just a cartographic nicety.

There is no such thing as too many maps for a prospector staking claims in the Canadian north. The entire validity of the claim depends upon accurate staking. It does not matter how carefully you pace off the perimeter of the claim if your first post is planted right in the middle of someone else's claim. And if you have made any error — no matter how small — in staking the property you want to claim, then another prospector can restake the ground for himself if he discovers your mistake.

The Province of Ontario is divided into nine mining divisions, each one presided over by a mining recorder. It is the duty of the mining recorder to receive and validate claims and to enter each claim upon the definitive set of township maps maintained by the mining division. These are the key staking maps. A staking map will comprise one township, and it will bear all the lastest staking data of which the mining recorder is aware. This does not mean that the staking maps are the absolute last word, for they are not. The absolute last word may be written on a post hammered into the ground out in the bush. A prospector in Ontario has a month to record his claim, and the staking maps will only carry the claims which the mining recorder has received. Still, the staking maps are the latest printed information available, and they are as up-to-date as a map can be.

Most professional prospectors are keen map collectors. Maps are their life's blood. They plot their staking strategies from maps, and they pore over maps the way a real estate speculator will run his eye over a likely row of buildings. There is so much gossip and rumour in the mining business that maps are always much more than simple co-ordinates to a prospector. Looking over an old staking map, a prospector's thinking might run like this:

"Yes, I know that township; I know that ground. Harry McIsaac told me that old Bill Knight staked that ground back in the thirties. He found a little silver, did a bit of trenching, but he couldn't option the property to anyone. The price of silver was too low. But it's not so low now. I wonder if that ground is open."

By checking the up-to-date staking maps in the mining recorder's office, the prospector can see whether the ground has been restaked, or whether the claim has lapsed and the ground is open. If he decides he wants to stake the property, chances are that he will assemble even more maps before he even sets foot in the bush. There are ordinance maps and topographical maps, aerial and ground surveys — anything that might show a logging road or a prominent feature of land, a mining work or an abandoned set of claims, anything that can provide a prospector with a point of origin is useful.

If a prospector's intention is to restake an old, lapsed claim, then the procedure is relatively simple. All he needs is to know roughly where the claim is located, walk in and tie on to the old claim stakes. Tying on is a term only. Prospectors do not actually lash new posts to the old ones. They simply use the location of the old claim stakes as locations for their new ones and describe the new claim as conforming to the data for the old claim registered on staking map such-and-such dated so-and-so. Sometimes there are as many as a dozen old, rotting claim posts lying around at the same location, for it is illegal to remove a post, even from a lapsed claim.

If a prospector is staking new ground in an unsurveyed township, then the procedure becomes more complicated. That is why the prospector needs all the maps he can get. The first thing he needs to determine is where to start his staking. For this he needs some readily identifiable feature to which he can tie on his first post. This could be a topographical feature, a mile marker on a logging road, a particular twist in a river marked by an outcropping, anything that is easily distinguishable.

The first post to be staked is at the northeast corner of the claim. Posts must be at least four feet high, with a smooth face cut at the top. This smooth face must be at least a foot long and four inches wide, and it must be cut on all four sides of the post, or slashed tree if a tree is used.

The information that a prospector is required to record on a staking post must always be written on the side of the post facing the next post on his claim. Thus, the information on post no. 1 must be written on the south face, facing post no. 2, which will mark the southeastern corner of the claim. The best instrument to use for marking the post is a felt-tipped marker that spreads ink. But if you are marking your post in midwinter, a marker using ink would be useless. The ink freezes. In the winter the best marker is a heavy pencil.

On the first post staked the prospector must write his name, his licence number, the date of the staking and the exact time at which he has started to stake the claim. The time is particularly important in a staking rush, when prospectors are scrabbling after the same ground and banging stakes in like pegs in a cribbage board. Licences for prospecting are available at the offices of any mining recorder in the province. In Ontario you must be eighteen years old and pay a fee of $50.00. A prospecting licence is good anywhere in the province, not just in the mining division where it was issued. It entitles the licencee to stake as many claims as he wants. But it is important to note that all prospecting licences in Ontario expire on 31 March and must be renewed before that date each year in order to hold and protect claims recorded under that licence. Let your licence slip, and your claims are lost.

On the three subsequent posts at the remaining corners of the claim, it is necessary only to record your name and licence number. The distance between each post is 1320 feet, and frequent marks should be blazed along the line of travel. Prospectors are also expected to thin out the underbrush along the blaze, so it is easy to walk from post to post. If there are no standing trees available to slash into pickets along the blaze, the prospector should mark his boundaries with mounds of earth or stone. Pickets or earth or stone markers should be placed at intervals of two chains (132 feet) along the boundary lines. The pickets are supposed to be 5 feet high, and the other markers 2 feet in diameter and 2 feet high.

If you are trying to stake ground that is partially submerged — that is, if a river or lake eats into a corner of your claim — then you must plant your posts on the boundary line at the water's edge facing towards the point where the post *would* go if the water were not there. Such a post is called a witness post and should bear the letters W. P. together with the distance to the submerged but true corner of the claim.

In addition to all of these requirements, a claim, to be valid, must also be tagged. Tags are made of metal, are about one inch by four inches and bear a number and the Ontario coat of arms. When you purchase tags from the mining recorder, the number on the tags is entered on a record next to your licence number. This is to make sure there is no confusion over whose tags are whose. It costs $1 for a set of four tags — one for each post — but the dollar is applied to the $10 charge for recording the claim.

There are some short cuts that professional prospectors use when they are in a hurry. Broadly speaking, professional prospectors are in a hurry at those times when other men are in a hurry, namely, when they suspect that the ground is loaded with gold. Once the boys on the diamond drills have started to outline an important discovery — oh, say a couple of million tons of gold-bearing ore grading 0.30 — then the gentlemen with the axes and the felt pens get a sudden urge to acquire ground. So does anyone else who can read *The Northern Miner*. This is when the temptation to pick up the pace a bit is more than human flesh can resist.

Let's say that two prospectors are working as partners, a common arrangement among professionals who discover that they are both interested in the same ground when that ground has suddenly become very hot to the touch. A partnership makes a lot of sense. For one thing, it means that instead of tearing around through the trees with a panting enemy on your heels, you are working the forest with a friend. For another, you are both increasing your chances of success. It is a rule of thumb in prospecting that for every 1000 claims a man stakes, 999 will turn out to have nothing on them but thin spruce and famished blackflies.

When two partners who know what they are doing begin a staking program in a hurry, they do not always carry along a copy of the Mining Act to make sure that every *i* is dotted and every *t* is crossed. Partners will usually parallel stake a property. It works like this:

Instead of each man staking several whole claims in a group of adjoining claims, each stakes just one half of the claim. Starting at a given point, probably the eastern boundary of the group they want to stake, the prospectors position themselves 1320 feet apart, or as near enough as they can reckon. This means, effectively, that one man is standing at post no. 1 at the northeast corner of the first claim to be staked, and the other is standing at post no. 2, the southeast corner. They begin to move west in a straight line, parallel. The second post to be staked will be, for the man on the north parallel, post no. 4 of the first claim and post no. 1 of the second, and for the man on the south parallel, post no. 3 of the first claim and post no. 2 of the second.

In this fashion two men can cover a lot of bush, and they can cover it fast. Now, at some point one of the partners is going to have to write the other partner's name on a post. You cannot have one prospector's name on the northern two posts and another's on the southern two. This will earn you nothing but a pitying smile from the mining recorder and an invitation to go back in and do it right. By the time you get back to your property, there could be so many new stakes around it that you will think someone is grazing cattle on the place.

One name only is allowed on any one claim and, strictly speaking, staking by proxy is illegal. And while you may have help in staking, it is the responsibility of the person making the claim to go round all the lines, place the proper markings on each post and record the claim at the divisional office. That is the short cut partners take. One man cuts the line along the northern boundary, staking as he goes; the other man stakes along the southern line. Later they share the work of cutting the north-south blazes that form the eastern and western boundaries of the claims. Just one man's name appears on the claim, of course, but they are partners in whatever the claim might produce. The advantages of this system are obvious. You get a lot of ground and you get it fast.

But the disadvantages are also obvious. Working fast, you can make a mistake. You make a mistake, you don't own that ground any more.

Staking a claim is one thing, holding it another.

The mere act of recording a mining claim does not confer ownership upon the holder of the claim. When a prospector stakes and records a claim, he has no title whatever in the land or minerals until either a patent, now rare, or more commonly a lease has been granted. All that a staked and recorded claim confers upon the holder is the right to *earn* the right to mine. Before such other rights are granted, certain assessment work must be done on the property, and verified by agents of the mining recorder. If the work is not done, the claim is cancelled.

A total of two hundred man-days of work must be done on each claim over a five-year period, taken from the time of the recording of the claim. This does not have to be manual labour, but it can be. Assessment work must be connected with mining, such as stripping away the overburden of soil and trees to expose the bedrock, or trenching the property. Prospecting is not assessment, and neither is the construction of roads or buildings on the property.

The Ontario Mining Act sets down exactly how much work must be performed in each of the five years: twenty man-days is required to hold the claim for the first year, forty man-days each for the second through the fourth years and sixty man-days for the fifth year. If more than the required work is performed on the claim in any one year, it is credited against the next year's quota. The Act also contains half a dozen specific ways in which man-day credits can be earned.

Similarly, if you own a large group of claims, as the major mining companies on the Hemlo camp all do, then you may apply work done on one claim to other claims which for the moment you are ignoring. Thus, the tremendous volume of labour being poured onto the key claims on the Hemlo camp will be spread around to help the claimholders maintain their rights on the whole block. There is a limit to this practice, however. No more than four thousand man-days completed on a single claim can be applied to other claims in the group. And all the claims in the group receiving credit from the heavily worked claim must be contiguous.

The following account is not included here to suggest in any way that the events occurred because of hurried staking. It is only to illustrate the perils of inexact staking, and the enormous amount of money that can ride on the accurate hammering of a length of wooden post into the thin overburden of the Canadian north.

The heart of the Lac Minerals property on the Hemlo camp is the Williams property. By the time Don McKinnon and John Larche were assembling their ground at Hemlo, the old blazed lines marking the perimeter of the Williams block had long since grown over. Now, Don and John knew those claims were there; anyone with a staking map would know that. But in the biting cold of winter, with snow on the ground, or in the first, blackfly-blown days of spring, or just in the normal push of getting the job done, they failed to locate the northeast post of the old, patented claim, and they staked inside the Williams property.

That piece of wood that bit ground inside the unseen boundaries of the Williams claim — that was one mighty important piece of wood.

The single stake that John Larche and Don McKinnon whacked into the Williams ground became a common post for three of the claims they were staking at the time. They had no idea that they were tying wedges of the ground they wanted to stake onto the patented claims, or they would certainly have moved their stake. Originally optioned by Hughes and Lang, the three claims involved — the three claims that became compromised by that single misplaced post — hold the richest ore on the Hemlo camp, the fabulous Golden Giant deposit. The mistake was discovered by Noranda — which had taken an option on the claims from Hughes and Lang — while in the process of restaking its claims to preclude exactly the kind of dispute that arose.

When a company like Noranda moves onto a camp, the first thing it does is resurvey the boundaries of the claims it has optioned. It does this to make sure that every single regulation and rule, every chapter of every statute, every paragraph that spells out what you have to do to own the right to mine a piece of ground, have all been followed to the letter. They do this in order to avoid the kind of unsettling situation where, having dropped $290 million into a hole in the rock, they find a pile of red-ribboned legal papers waved in their face, and a sheriff telling them to shove off. And so, unless you are so dense that you ought to be pounding your staking posts into the ground with your head, you will take every precaution to protect your interests. One of these is to survey.

What Noranda found out when it went in to resurvey The Golden Giant made a lot of people back at headquarters uneasy. Noranda found that post and discovered that it was on somebody else's claims.

Noranda quickly restaked the property in a way it hoped would avoid a dispute with Lac, which owned the option on the Williams patents. It was a vain hope. One of the little wedges of property involved in the dispute, a tiny strip about 120 feet wide and half a claim long, was estimated to contain anywhere from 250,000 to 400,000 tons of high-grade ore. Worth? Perhaps $4 million to $6 million.

Certainly worth an argument.

Lac responded swiftly. It struck into the bush and restaked the entire three claims. In effect, Lac was claiming The Golden Giant. The Hemlo camp went wild, and charges of claim jumping flew like snowflakes in a blizzard.

Audrey Hayes, the mining recorder at Thunder Bay, began to feel the pressure. When Noranda found that one of its posts was on Lac ground, it had asked the mining recorder for an opinion on how to proceed with the restaking in order to head off a dispute. Audrey Hayes, together with Noranda officials, agreed that Noranda should restake the property post-for-post, meaning that Noranda was to stake its posts beside those of the original McKinnon-Larche staking, even though at least one of them was on patented ground.

But Lac disagreed and felt that the restaking should have been along the newly surveyed boundary. Lac felt that the recorder's decision was wrong, open to dispute, and that therefore the whole block of three claims was open. Ian Hamilton, a lawyer for Lac, said at the time that the whole episode had been "the kind of dispute that arises often in an area where there has been a lot of staking, some of it good, some of it sloppy."

The two mining giants eventually settled the dispute between themselves, with Noranda holding The Golden Giant and Lac getting a thin slice of a very rich pie. It is for reasons such as this that the Ontario government prints at the bottom of its summary of staking requirements the advice: *Do Not Hurry*.

Ten

An Ordinary Multimillionaire

Leave Murray Pezim's golden tower, turn left, walk a couple of blocks down the hill and you are in another world. This is West Hastings Street. West Hastings Street is the Bank of Canada. West Hastings Street is the Vancouver Club and the University Club. This is the old financial street, the old address, before the boom and the tinsel surge carried them all uphill, south, towards the new stock exchange and the bursting hotels and the shiny new office towers.

West Hastings Street is the old sanity of Vancouver, where the city parked its money right down by the water to keep an eye on the timber and the shipping, the sources of all that wealth. There is a stately edifice at the corner of Burrard Street and West Hastings, the Marine Building, that could be the most beautiful office structure in the country. An extravagant frieze above the main storey is studded with the thrusting prows of sailing ships, adance with the play of dolphins, full of fancy and artistry and care. The building says: I am here to stay.

No. 675 West Hastings Street is a discreet address. No grand pavilion elbows forth into the street to capture the pedestrian with its glossy allure. No. 675 West Hastings stands as straight as a dowager dressed for court. This is a building that has arrived, and the struggle of those still trying to get there is a spectacle worthy only of contempt.

Inside the front door, the lobby is all gleaming marble and polished brass. The elevator doors shine like the portals of a temple. Off in a corner there is a little brass railing. This fences off the command position of the chief porter, a ramrod of intimidation and glaring aloofness who scrutinizes the human traffic as if he were confirming an awful suspicion that he has been harbouring all his life. No one really relaxes here until the elevator doors close on his baleful regard, and the collective sigh is palpable.

The interiors of these old cars are finished in a satiny wood that glows in the soft light of brass sconces. The doors open and close on dignified, simple hallways, utterly anonymous. No one getting on or off calls anyone else "buddy." But do not be deceived. These are not insurance agents. On some of these floors, along some of those simple, anonymous hallways, are doors that conceal some very, very busy people, people who own the rights to dig for gold.

The fifteenth floor is right at the top. When you enter the reception area, the first thing you see is a large map of Canada that takes up the whole wall behind the receptionist. Above this map is the legend, The Hughes-Lang Group, and the map itself has enough points marked off to show that the Hughes-Lang Group does not spend its time hanging around over the same couple of holes in the ground from one week to the next. Up here on the fifteenth floor, if the dice are going to be rolled with Mother Nature, then by God there are going to be more than one or two rolls.

The room is carpeted in a handsome gold, and the walls defined by those heavy, dark wood mouldings that tell you the building was built by carpenters, not just popped together like pieces of Lego. You are here to see Frank Lang, one of the partners, and you are invited to sit and browse through the reading material while you wait. The reading material is principally about the Hughes-Lang Group. There are annual reports for some of the companies, and there is a mining newspaper or two. There is no *Time* or *Newsweek*. If you want to know what's going on in the world, go to the dentist. No one offers you a coffee up here in the sedate eyrie of these two well-tried eagles. The Hughes-Lang Group is not a coffee shop.

Frank Lang does not look like a mining promoter. A promoter is supposed to be full of crack and sizzle, swish and pop. In fact, you expect any promoter to look a little like Murray Pezim, since it's The Pez who has stamped the promoter's image so thoroughly on the whole Hemlo event. The Pez has made it his own, as far as the media is concerned. But The Pez doesn't own all the ground. Oh no. The gentlemen up here on the fifteenth floor of this singularly discreet grey stone office tower on staid old West Hastings Street, they own plenty.

Frank Lang is around sixty years old. He is wearing a rather ordinary grey suit, a pale blue shirt and a grey knit tie. His shoes are black, laced, and have that pebbly finish. Frank Lang does not look at all like the kind of promoter who would give the people on Bay Street fits of apoplexy on sight. No, indeed. Frank Lang looks like the kind of man who might be working for the pin-striped aristocrats at Bay and King. Frank looks exactly like the kind of man about whom one of the lordly folk in Toronto might say, "Frank'll get it for you." But Frank wouldn't get it for you. Frank would tell you to get it for yourself. For even if you are one of the mandarins who populate the imperial offices at Bay and King, the chances are very good that Frank has more money than you. A lot more.

Frank's appearance suggests the bureaucrat. And for thirty years Frank *was* a bureaucrat. As soon as he took his degree from the University of British Columbia — major in physics, minor in mechanical engineering — Frank went to work for the old Department of Mines and Technical Surveys in Ottawa. This may be where Frank learned how to dress. In 1955 he quit his job in Ottawa and came back to British Columbia, where he took a job as an engineer with B. C. Hydro. That is where he stayed until he retired a few years ago after twenty-five years of faithful service. He even has a gold watch. Thing is, that's not the only gold that Frank owns. Two of the pins in that big map behind his receptionist's desk are stuck into the Province of Ontario. At Hemlo.

While other men and women simply headed home after a day in the office, maybe went down to the basement to knock together a birdhouse for the back yard, Frank would go home, head down to the basement and knock together a company. A company, if you're clever, you can sell to someone else down the road. A birdhouse — well, go ahead and stick it up and see for yourself what the birds leave you at the end of the year.

Frank Lang beholds the world through a pair of friendly blue eyes. He smiles a lot but watches you very closely, as if you were a piece of intriguing rock that at any moment might reveal something worth mining. Probably you will not, but he'll keep a sharp eye out just in case.

"I started prospecting with a friend of mine back in Ottawa when I was still working for Mines and Tech Surveys," Frank recalls, smiling at the memory as he smiles at everything. "We used to go out on weekends, go out into the country and just look for rock. Then Blind River happened."

Blind River was the great uranium strike of 1954. It was the Blind River strike that built Elliot Lake, Ontario. Blind River made a lot of fortunes. Blind River sent countless men from every corner of the world — Finns and Germans and Poles and Scots — pouring into the miserable bush to find illness and death and heartbreak and hundreds upon hundreds of miles of worthless, empty, soul-smashing rock. And uranium.

"We got caught up in Blind River ourselves — who wouldn't? — and we went up there and did a little staking. Just on the fringes, but you didn't have to be sitting right on top of the stuff to make money."

Frank smiles. "I made $300."

But it was enough to whet his appetite, and when he returned to British Columbia the next year to work for B. C. Hydro, his eyes were on the bush. He had taken some geology courses at Ottawa's Carleton University and was ready to cock a more critical eye at the rock. It was shortly after he returned to British Columbia that he met Dick Hughes and formed the association that has planted their names at the head of a list of companies that is known to investors as far afield as Europe.

"It was just an investment club that we started in 1958. We were both working for Hydro. Mining was dead from 1955 until 1962 or 1963. Dick and I, we were just casual investors, nothing more than that."

And the Bank of America is just a bank.

Lang's office takes up a corner of the building. The windows are tall and many-paned, with elegant curved handles. If you want to open the windows in this fine old building, you can do so without heaving your desk through them. One wall faces south towards the stock exchange. The other set of windows looks west towards the gold-brown sheath of the Daon Tower and beyond. Below, a massive restoration of the old federal buildings is underway. These are handsome, venerable structures. More importantly, they are *low* structures. Frank Lang's view is never going to be obstructed by someone else's building. Between two tall windows, one framed photograph stands out, so arresting, so beautiful, that you are drawn to it across the room, approaching it with eyes that widen as the detail becomes clearer.

The photographer has chosen a piece of crimson velvet as the background for his subject. Upon this background the objects of contemplation stand like members of some stiff and stylized royalty: two pieces of core pulled from the rock by a diamond drill, a broken piece from each core lying before it on the velvet.

The cores are mostly quartz, and they glow. The photographer has lit his subject with devotion, not so it sparkles at you like tinsel, but simply states its own importance. Within the quartz are dark veins of some granitic intrusion. These are interesting only as a counterpoint to the beauty of the quartz. And the gold.

The gold in these cores is so astonishingly rich that if you could pluck it out of the quartz, you could hammer it into a necklace right there on the spot and walk away with it warming your skin like a Pharaoh's collar. So rich is the vein struck by the drill that pulled this core that when they plunked it down on the assayer's table, you could hear the sound in Johannesburg and Moscow. A lot of envious and unfriendly people who make their homes in other lands know about this core, and a lot of them do not like it. This is the kind of core that makes people with other gold mines have bad dreams in the middle of the night and wake with dreadful thoughts. And this is why:

Some of the richest gold properties around Hemlo grade 0.25 ounces of gold per ton. The gold in the core in the photograph on Frank Lang's wall, from one of his Quebec properties, grades *3677.8* ounces of gold per ton.

Put it another way: Let us say that gold is worth $400 an ounce. Then to make a million dollars' worth with ore grading 0.25, you need to dig up *10,000 tons of rock*. But if your gold grades 3677.8 ounces per ton, then all you need to dig up is *three-quarters of a ton*. For heaven's sake, you could hack the stuff out of there with a pickax while Noranda was mortgaging the farm to sink *its* mine, and you'd be ten times richer in half the time. *If* there were enough. This same photograph appears on the front cover of the 1980 annual report for D'Or Val Mines, one of the Hughes-Lang Group. On the inside of the cover the following caption appears:

"Considerable amounts of free gold appear in the diamond drill core from Hole No. 29 of D'Or Val Mines Ltd.'s Beacon property. While it assayed at 3667.8 ounces of gold per ton, this spectacular sample is not considered representative of overall grade. Plans have been made to donate it to a mining museum in Quebec."

"Not considered representative of overall grade" is an understatement that ought to have its place in the *Guinness Book of Records*. If that core were representative of grade, you wouldn't be able to keep the claim jumpers off with a Tommy gun. People would be tunnelling towards that deposit from the Montréal Métro, never mind how long it took to get there.

And here is where you catch the first glimpse of the promoter. Lang is the first person to agree that this is not representative of what they have *so far determined is there*. But he is smiling as he watches you, closely, and adds:

"But you never know what's down there, do you? You never know until you drill. This is not like Hemlo, where we know the gold is spread through the whole ore body. In quartz, the gold is in veins. You could miss that by a quarter of a millimetre, you'd never find it. How many more are there, that's all I ask. How much more gold, just as rich as that, is down there?"

Smile. Watch.

"Only way to find out is drill. Just drill."

This is pure promoter. They catch you staring at this core loaded with gold; they catch the prospect as he stares at the glistering fatness of the vein, drifting down there through the quartz, and they whisper: "What if?"

Of course, it isn't just some pretty tubes of rock bright with promise that make you listen to Frank Lang. After all, Frank is one of the Hemlo millionaires, one of the men who were smart enough and fast enough to see it, grab it and hold it. He and his partner Dick Hughes optioned 156 claims from Larche and McKinnon in 1981. That Frank was a rich man before Hemlo, no one will argue; but Hemlo is the gold bar winking at the back of everyone's brain, the signal of something fabulous. To have grasped some of it is to become more than just a lot richer. It is to become tinctured a little with prescience. It is magic.

Lunch with Frank Lang is not as rarefied an experience in the pleasures of the table as lunch with a multimillionaire ought to be. We drift aimlessly out onto West Hastings Street, and Lang seems a little uncertain about which direction he wants to head in. If you have been expecting a swing to the right and a short march down the street to the Vancouver Club, you can forget it. Instead we turn left, idling along the sidewalk for perhaps a hundred feet. We pause beside a flight of concrete steps leading up to the second floor.

"They have a section in here, I think, where the waitress comes to your table to take the order."

We ascend. There is no such section, so we just plunge into the line. Hell, who needs a lot of stewards in scarlet cutaways? Let's just have some more of that potato salad.

Trays loaded, we make our way to the rear of the cafeteria. This is a window seat at least, but the view does not exactly take your breath away. Six feet in front of the window workmen stagger around under the weight of buckets in the gloomy interior of the building they are erecting between us and Grouse Mountain. We look at them. They look at us. We turn back to our food. Lang talks about Hemlo.

"In 1980 we took a verbal option from Don McKinnon which was to include the whole thing. Everything. All the property they'd staked — fourteen claims so far. Mind you it was only a verbal agreement, no money changed hands, but it was still an agreement. This is a handshake business. A handshake in the bush and you've got a deal.

"In Christmas of 1980 Don McKinnon — I guess he needed money — called me and asked if he could sell the property they'd staked up to now to someone else. Well, we only had a verbal agreement. I said, sure, okay, to whom? Don told me it was Nell Dragovan, and I said sure, go ahead, Nell's a nice girl."

Lang stops to watch the workmen for a moment, putting his fork down and taking a sip of tea. Then he looks at you and smiles.

"I didn't know Nell was connected with Murray."

Lang does not make a big thing out of this. He does not scowl and he does not criticize. He does not even suggest there was the tiniest scrap of deceit. All he says is that he didn't know "Nell" meant "Nell and Murray." But he probably would have let Don move the ground anyway, because Lang and Dick Hughes were not having much luck all by themselves.

"We were going up and down the street trying to sell Golden Sceptre and Goliath [the two Hughes-Lang companies that ultimately optioned Hemlo]. We were trying to sell those companies to anyone who would buy them at seventy-five cents a share."

At the time that Lang is recounting this tale, his stock is selling at $21 a share. He pauses to point this out. It is pleasant to be able to record, very plainly, for all those people who would not buy at seventy-five cents, that, had they bought, their investment would have increased by a factor of twenty-eight. Had they bought $1,000 worth, they would now have $28,000 worth of stock. And so on. Lang pauses so that the next words will emerge all by themselves, uncluttered, stark.

"We were fortunate enough to run into some Rothschilds people."

It is strange to hear this emerald name dropped like a forkful of coleslaw into the conversation of two men who have carried their own trays to the table. Lang, here, was fortunate enough to run into some

Rothschilds people, and he just feels you might like to know. These are the people to whom maharajahs come when their princely states are collapsing and they need to convert unwieldy property into cash. Some Rothschilds people would know how to help them. Rothschilds have helped kings and crooks for several centuries, somehow managing to avoid losing the whole bundle in the wars that slop across Europe from time to time, erasing less agile families as though they had never been.

And so to Frank Lang and Dick Hughes, two men with a gold property that turned into a stunner, came the fortune of an encounter with the Rothschilds. Who bumped into whom, you may wonder for yourself.

"The Rothschilds kind of broke the back of the issue for us. They took about 30,000 shares of the 200,000 we had to offer. They saved us. We were on the verge of losing those claims. We almost did lose them.

"We were nip and tuck to hold onto that property, and they really saved our bacon. We appreciated it so much that we gave them a private placement at $4.50 a share before the deal with Noranda, and by the time the ink was dry on the Noranda deal those shares were worth $20 each."

It was the Rothschild money that kept the Hughes-Lang Group alive at Hemlo until the deal with Noranda. At the time, Bay Street was still slumbering in its red leather chair, unwilling to grant to Hemlo any more importance than a little noise in the bush from a bunch of bozos from Vancouver.

The partners directed that a series of holes be drilled along an east-west line approximately five hundred feet north of the boundary between their own claims and the Lac property. They believed that the Lac mineral zone extended into their own claims, and that if they drilled they would intersect it at a depth of one thousand feet.

The first hole intersected a geological fault, and they abandoned it.

The second hole found The Golden Giant.

When they pulled the core on the second hole, they measured 98 feet of mineralization with an average grade of 0.256 ounces of gold per ton. The gold plunged to a depth of 1070 feet.

"I was so confident after that that I took off for a walking trip in to Mount Everest," Lang recalls. "I expected that when I got back, the stock would be trading at about $15. But it was at $5." Lang shakes his head. "We were pulling out some pretty good holes, grading 70 or

80 feet at 0.30 ounces per ton. It didn't make any sense. It made no sense at all.

"Then when I got back I heard it. There was a rumour that we'd been salting our core."

Salting a core means falsifying the core, adding gold that was not there when you pulled the core out of the ground. You can salt a core by slashing a gold ring along the side of the smooth rock cylinder. *Some* gold will come off the ring and stick to the core. That is a simple method. It is also not all that hard to detect. But there are a lot of cagey brains in the world, and plenty of ways to salt a core that even people trained to look for it cannot spot.

But rumour could only slow the inevitable. The great gold-laden slab of rock that became known as The Golden Giant deposit was there, and the partners kept poking holes into it. The initial drilling program consisted of another nine holes. Those holes indicated potential reserves of 6.5 million tons of ore averaging 0.25 ounces of gold per ton. Nobody salts with that size shaker, and Noranda went for The Golden Giant like a lonely lover. The mighty conglomerate optioned the Hughes-Lang Hemlo claims. All of a sudden, nobody was sleeping on Bay Street. The big kid on the block had finally joined the gang, and stock in Golden Sceptre and Goliath went up so fast you had to be holding on with both hands.

Noranda rebated to Hughes and Lang the $750,000 that the two promoters had spent proving The Golden Giant up to the time Noranda joined them on the camp. That was the *small* change. The real price for Noranda's entry at Hemlo was a massive purchase of stock in the two Hughes-Lang companies operating on the gold camp. At $6 a share, Noranda came in for 400,000 shares of Goliath and 600,000 shares of Golden Sceptre. Six million dollars. Not bad for a down payment.

The Rothschilds, reportedly, are content.

Eleven

Alf & Co.

Just about everything that God makes without the intercession of man, Alf Powis sells. Except seawater. Alf doesn't sell seawater. Yet. But if a time should ever come when a market for seawater develops — maybe the Russians will find a way of turning it into wheat — then Alf will be into that market so fast you'll think seawater was a Canadian invention.

Alf Powis is the chairman and chief executive officer of Noranda Inc., a breathtaking natural resources empire that spins its awesome web of products out of the very stuff of planet Earth. Copper. Zinc. Aluminium. Lead. Silver. Molybdenum. Potash. Paper. Lumber. Oil. Gas.

And gold. Noranda sells gold, too. Gold is what brought Noranda to The Hemlo. Crippled by the global collapse of its markets, recession, and an exhausting struggle for control of its own board, Noranda needed good news and needed it fast.

When Noranda came, so did faith. Noranda swept onto the gold camp like an empress, and the rest of the place just stared in wonder as she brushed the wilderness aside and started to dig. Watching Noranda go after the gold of Hemlo is spellbinding. She rushes about in a fury, imperious, stubborn, determined and magisterial, ordering the affairs of her people and pouring money onto the site like a profligate consul wielding the power of Rome. Like Rome, she is powerful. Like Rome, she looks to outsiders like a seamless entity of uniform will.

Like Rome, she boils at the core.

Noranda is the enchanted child of the Canadian mining industry, born back in the mythical days when Noah Timmins was still running Hollinger Mines in Ontario. Everything coloured by association with Hollinger is enchanted in the lore of Canadian mining. Until they struck The Hemlo, Hollinger was the richest gold deposit ever in the history of the Americas, a fabulous lode that drew men up to the northern woods as if it had been Babylon. The vision of Hollinger, glittering along her quartz veins like some ferocious, wilful regent shut into a palace of solid rock, is the vision that drew men into the lonely bush to chip away their lives in a world of endless granite.

The men who started Noranda were looking for Babylon, too. But instead of drumming through whatever parts of the Ontario bush had been ignored by Noah Timmins, they decided to strike across the border into northwestern Quebec, where prospectors had staked interesting new formations. Geologists like to find places where the formation of the rock suddenly changes. It is these geological anomalies that indicate the rock may be mineralized, that is, contain something more valuable than the raw material for a good fireplace.

In 1922 two American financiers picked up the claims staked on the Quebec geology, and with $50,000 formed a company to drill the eight-hundred-acre site. The results were disappointing, and it was only to fulfil the exact terms of their contract that the diamond drillers even bothered with the last hole. Pay dirt. They intersected a layer of mineralized rhyolite 131 feet thick. There was no gold worth mentioning in the core they pulled, but the copper values were staggeringly rich. And then a strange thing happened.

The Americans commissioned an engineer to come up from New York City and examine their exploratory shaft to see if it was worth building a smelter. The engineer made the long journey north, studied the site, read all the reports and poked around in the shaft. His conclusion: "I wouldn't piss on it." There is no point in contemplating the folly of this unfortunate New Yorker. He was simply a man who was asked for his opinion and gave it. One hopes that his children never found out about the chance he blew, and that the obscurity of his declining years were not too troubled by remorse.

Within a year the Americans sold out. Canadians bought in, among them Noah Timmins, who advanced the $3 million needed to build the smelter. What developed was a copper deposit so rich that an entire city grew above it.

And what developed was Noranda Inc.

Noranda has assets of more than $6 billion. There are not too many organizations on earth that own that much, and certainly there are plenty of small countries whose problems would be solved for a very long time if they had the assets of Noranda. Their problems would probably be solved if they had the *management* of Noranda, but never mind. Some countries have better things to spend their money on than managers, such as palaces for their life presidents and lively, futile little wars.

The fact that Noranda is a big company has not exactly escaped the notice of the Canadian business community. Noranda is ranked twenty-seventh on the 1984 *Financial Post* index of the top five hundred companies, and you do not get near the top of the ladder without having a lot of the people lower down staring longingly at your rung. But that's okay. You just scrape a little sand off the bottom of your shoe, so that it falls down into their eyes, and smile. They are no real threat to you and your tight gang of managers, dug in behind your piles of minerals, your oil and gas and your trees. No one is a threat to you.

Unless he has a few billion dollars and wants to buy your company.

Noranda is a public corporation whose stock is widely held. It was even more widely held back in 1979, before two cash-heavy gentlemen named Edward and Peter Bronfman decided they would like to acquire a position in the company. When people acquire a position, they usually acquire a voice in how the company is going to be run. This is what the Bronfmans' company, Brascan, wanted. And this is what Alf Powis and his management team *didn't* want them to have.

War.

The battle for Noranda began because the company's management had run the company so well. Until the recent savage downturn caused by a world recession, Noranda's return on investors' capital was the highest of any of the country's top thirty corporations. God made the minerals, down at Noranda, and the boys on the smelter just boxed 'em up for delivery. The wonderful thing about an arrangement like this is that God will not jump up and down and demand a seat on your board of directors. You get the Bronfman boys in there and, bet on it, they're bound to be less benign. So ran the thinking on the forty-fifth floor of Commerce Court West, Toronto, where Noranda's field commanders keep the keys to the gunpowder.

The building is the Bank of Commerce's contribution to the tower garden planted at the corner of King and Bay streets, the heart of Toronto's financial district. The forty-fifth floor of Commerce Court West is a powerful address.

The preliminary skirmishing was over the block of shares that had fallen into the ambit of Argus Corporation, the elegant corporate tool that E. P. Taylor and, later, financier Bud McDougald used to erect their images on the altar of Canadian finance. Argus dwells like a regal serpent coiled up within its polished little neoclassical jewel of a building at 10 Toronto Street. There, behind the great, gleaming black doors that hold the entrance firmly shut against the prying gaze of the shabby herd, the pin-striped lords of Argus manipulate their figures, buy and sell, make and break.

Argus's shares in Noranda came down to it through its ownership of the fabled lady of the north, Hollinger Mines. Hollinger owned Noah Timmins's original block of Noranda stock. Originally a holding of 200,000 shares, the Hollinger block had grown through subsequent stock splits until it comprised 11 per cent of Noranda. Back in 1976, when Bud McDougald was still alive and Conrad Black, perforce, was still chafing and straining and burning away in the shadow of the boss, Black suggested that Noranda stock was undervalued and that the company was a very ripe piece of fruit that Argus would do well to pluck.

What Black was saying, in effect, was: "Bud, let's buy the sucker." What Bud said was: "No."

But Conrad Black did not get where he is, which is where Bud McDougald *was*, by saying "Yeah, I guess you're right" every time someone disagreed with him. No, Conrad Black has a memory like a building full of Univacs; once he gets an idea in there the only way to dislodge it is to buy it. And Conrad Black had the idea that it might be fun to take a run at Noranda.

What Black did was methodically begin to assemble shareholders lists. He wanted to pinpoint the people with large enough blocks of shares to make it worthwhile approaching them with more tenderness than a clenched fist. What he would say to them is anybody's guess, but Conrad Black is known to be a man capable of waxing rather eloquent when the subject is proprietorship. What he says is: "It's good for rich guys to own things, 'cause we care for them." This is an arguable premise, of course, but it is generally received with unequivocal support when it is being pitched to other rich people. How they would receive it in Moscow is another matter.

Back at Noranda, Alf Powis got wind of Black's machinations. He did not like them. This is why:

Let us say that an investor with access to an enormous pool of capital decides that a company listed on the stock exchange is trading at a price below what the shares are actually worth. In other words, the shares are being bought and sold at a price below what they would fetch if the company's assets were all liquidated, its obligations met, and the net cash spread among the shareholders. If someone with fiscal clout decides that is the case, then he might be tempted to buy the company cheap on the market and sell off its assets, paying the proceeds back to himself as dividends. This is called stripping a company, and what is left when the profit-takers are through is a building with a For Sale sign on it and a lot of former employees looking for work. Whether or not this is the scenario perceived by Powis, it is at least a possible scenario, and the chairman of a corporation the size of Noranda does not get where he is by ignoring possibilities.

Powis called in his emissaries and gave them the word. They were to approach Conrad Black and offer to buy Hollinger's 11 per cent stake in Noranda. The offer was for $37 a share. Black's response was terse. He told Powis he was nuts if he thought Black would let his stock go for $37 a share. Powis countered by offering Black a seat on the Noranda board, reasoning that making Black an insider would head off any deals inimical to the interests of Noranda as perceived by Alf Powis. Black said he would love a seat on the board, as long as he could raise his stake in the company to 20 per cent.

Powis got the message and move swiftly.

At Powis's command, Noranda's treasury issued four million new shares at once, using the new stock to purchase the full assets of a subsidiary company, the zinc producer Mattagami Lake Mines. Under the tidal wave of new Noranda stock, the Argus block became diluted, losing some of its punch. Powis wasn't through. He ordered Noranda stock to be split three-for-one. The move further diluted the Argus position. Conrad Black said to hell with it. ''I thought we might win if we went after Noranda,'' Black told author Peter C. Newman during interviews for *The Canadian Establishment,* ''but it would have been a real corporate Vietnam, with Hollinger pouring more and more resources into the battle and Powis issuing more and more stock. That would have been very messy, expensive — and nobody could guarantee a final verdict.''

Black had thrown in the towel as far as his own lunge at Noranda was concerned, but he had decided that one way or another he was going to make money out of that stock, and if he couldn't turn a profit by buying more, then he'd turn a profit by selling what he had. Black let the world — or that part of it with the cash — know that the Hollinger block of Noranda was for sale. The nose that poked out of the woods for a sniff was attached to the face of Trevor Eyton, riding point for the Bronfmans. Eyton had decided that Brascan could use a nice company like Noranda and offered Black what amounted to $64.50 for each one of his 7,850,490 shares. Black took the deal, and the very day that the papers were signed Brascan hit the market, looking to beef up its position in Noranda with the acquisition on the open exchange of another million shares.

The fight for Noranda was on in earnest, and the company's executives closed ranks around the chairman.

The pride that Noranda's executives take in their company is evident as soon as you get off the elevator at Commerce Court. One wall of the huge lobby is covered in a great sheet of copper, worked into an intricate mural bas-relief that depicts the history of the company. The sweep of windows looks west, and on a clear day you can see the shore of Lake Ontario bend around as the Niagara Peninsula strikes across the top of the lake to the American border at Niagara Falls.

Polished shafts set in a cluster fill one corner of the room. Each shaft supports a glass case, and within each case is a large chunk of ore. Copper. Zinc. Molybdenum. Gold. The foundations of an empire.

Alf Powis is an exceedingly courteous man, even courtly. His expression is warm and his smile ready. In the lapel of his suit he wears the pin of the Order of Canada. His hair is curly and inclined to pile up in a wave. The overall picture is of a man at once dignified and friendly. And in fact this is the consensus on the street. Among the aggressive, gossipy, uncharitable pack of business journalists who follow the movements of men like Powis, the widely held opinion is that he is simply a nice guy. No doubt this is true as far as it goes, but it would be wrong to conclude that Alf Powis is a teddy bear. Teddy bears do not rise by the time they are fifty-two to control companies that employ 55,000 people. Teddy bears are not invited to join the Toronto Club, that communion of saints where people like Conrad Black meet to divide up the country as if it were personal property.

Alf Powis began his career at Sun Life, moving into the investment department straight out of McGill University when he graduated in 1951. Sun Life stood at the heart of English Montreal in those days, when English Montreal stood at the financial heart of Canada. There were richer institutions, certainly, but none housed more gloriously than Sun Life in its colossal palace of stone on Dominion Square. Sun Life was the symbol of English lordship in Quebec, and that they have not torn the place apart just for the sheer joy of it is a testament to French forbearance. The investment department of the old aristocrat of a company was a good place to learn how Canada ran, and Alf Powis learned.

Powis's chief lieutenant is Adam Zimmerman, Noranda's president and chief operating officer. Zimmerman is a little older than Powis and looks more exactly like what he is. No one would ever mistake Adam Zimmerman for a teddy bear. He is tall and lean and hard. And no matter where you cut Adam Zimmerman, the blood that comes out will be blue.

Zimmerman was educated at Upper Canada College, Ridley, the Royal Canadian Naval College and the University of Toronto's elite Trinity College. His clubs are the York, University, Mount Royal, Toronto Golf, Craigleith Ski and Madawaska. None of these clubs hands out membership applications on the Toronto subway. These memberships, eighteen directorships and a very sharp mind all conspire to keep Zimmerman right where he is: at the top.

This is the formidable partnership that closeted its senior officers in the big suite on the forty-fifth floor of Commerce Court West and plotted the defence of Noranda from the Bronfmans.

The Bronfman pursuit of Noranda was relentless. They kept buying stock, buying stock and buying more stock. The Brascan offices are also in Commerce Court West, three floors above Noranda's. Trevor Eyton got very used to taking that three-floor ride down to the offices of his reluctant quarry. There he would repeat his assurances: Brascan admired Noranda's management. They thought Noranda was a brilliantly managed company. That is why they wanted to acquire a position in Noranda. In fact, they had already acquired a position in Noranda, and now could they please have a seat on the board?

The first time Trevor Eyton took that short elevator ride, Alf Powis was out of town inspecting one of Noranda's properties in Alberta. Adam Zimmerman was guarding the fort, and Adam Zimmerman is the man whom Trevor Eyton went in to see. Zimmerman is a cool fellow. There is no falling off the chair when Zimmerman is receiving news, pleasant or otherwise. These might be the guys who knocked off Brascan when Brascan got so fat it couldn't carry its own weight, but, by God, they were going to find that Noranda could take a punch and not even sway from the blow. This is not what Adam Zimmerman told Trevor Eyton, though. Zimmerman just told him he had great taste in companies.

It was a punch-up. The Bronfmans continued to build their stake in the company, and Noranda's executives continued to steer their company along a tricky course that involved the kind of fiscal legerdemain only the most adroit and nerveless practitioners can execute. All of these fervid, if deft, gymnastics served to pose a central question to the curious and rapt bystanders who watched the battle unfold day by bloodstained day. The question that Bay Street was asking was: Why? Why fight the Bronfmans?

Noranda is a public company, its shares traded on the open exchange and open to purchase by anyone. Its executives are not owners. The company is not *their* company in the proprietary sense. Their stake in the company is professional rather than dynastic; there is not some personal fief to be preserved, increased and passed along to an overfed band of little princes learning to play cricket at Upper Canada College. What the running of Noranda is supposed to be is a job. You come in in the morning, you do your best for the shareholders, you go home at night. If you do well, then the directors see to it that you are paid a crillion dollars for your efforts. At age sixty-five you clean out your desk and prepare to spend the autumn of your life between Georgian Bay and Lyford Cay in the Bahamas.

Simple?

Not simple.

When a pair of managers as dedicated, as imaginative, energetic and creative as Alf Powis and Adam Zimmerman get their hands on a company like Noranda, that company becomes their company as far as they are concerned. They pour their souls into the company; they make it grow and thrive and cast its roots into a hundred enterprises. They plot and manipulate and execute and direct. What they will *never* do, people like this, is turn placidly away while people whom they do not really know very well come in and buy their company.

By all accounts Trevor Eyton spent a lot of time assuring the Noranda management of his good and honourable intentions. He spoke, if the reports are true, of the position of his employers, the Bronfmans, as simply that of responsible investors interested in taking their rightful place on the board of a company in which they owned a substantial interest. They did not want to strip Noranda, Eyton protested. We did not come in here, he was saying, to ruin the glorious construction that you and your fellows have so masterfully wrought. We have come in here because we like what you are doing and want to share in it as responsible colleagues.

The negotiations and pleadings, the arguments and the wooing, continued. Powis drove north into the Caledon Hills one weekend for a particularly candid confrontation at Eyton's country estate. Eyton invited Powis, another time, for a cozy dinner at Stop 33, the swank supper club perched atop the Sutton Place Hotel. Eyton pleaded the Bronfmans' case, the case of the responsible investor who wants only to share in the benefits of the wonderful management already in place. Powis remained elusive. The Bronfman fear was that Powis would attempt to dilute the significance of their holding in Noranda by pulling the same lever he'd pulled against Black: issuing more stock out of the company treasury.

And he did.

Powis issued $100 million worth of treasury shares. Eyton offered to buy. But Powis was already well informed about Eyton's vigorous buying campaign on the exchange and was concerned that if he sold any of the new issue to Brascan it would be simply selling them the company. Instead, he planned further moves.

The entire structure of Noranda's ownership — how it was held by subsidiaries, what it owned, what they owned — was restructured in a complicated series of transactions that resulted in Noranda buying some of itself. Several Noranda subsidiaries clubbed together to form a company called Frenswick Holdings Limited, which in turn joined with still other Noranda subsidiaries to form yet another company called Zinor Holdings Limited. Zinor, by receiving a huge block of Noranda treasury shares and through a $266-million loan from the Canadian Imperial Bank of Commerce, where Powis sits on the board of directors, was able to end up owning *23.6 per cent* of Noranda Mines.

The move was a terrible blow to Eyton and the Bronfmans, and they felt it keenly. In one swift and elaborate manoeuvre, Alf Powis had slashed their holding from 16.3 per cent of Noranda to 14 per cent. If this doesn't sound like a very big deal to you, then try dropping $138 million down the pipe yourself and see how it feels.

By now, Eyton and the Bronfmans are getting even more edgy. Their big fear is that Alf Powis might do something which will water down the value of the Brascan block of Noranda shares even further. They are right to harbour this fear, for that is what is happening. Powis has decided to buy the MacLaren Power and Paper Company of Buckingham, Quebec. He is going to pay for it — you guessed it — with Noranda treasury shares, 10.4 million of them. The effect of this on the Brascan block of Noranda stock will be to reduce it from its already diluted 14 per cent to 12.6 per cent of the company. The drop from 16.3 per cent to 12.6 per cent is a paper loss of $222 million.

Even for rich men, $222 million is a lot of lolly.

The Bronfman interests are beginning to feel unloved, and some very determined men begin to sit down at board tables, roll up their sleeves and decide what they are going to do about it. They are beginning to get nasty-minded, these men. They feel as if, having shined their shoes, bought the box of chocolates and tied on their bow ties, the date has just slammed the door in their face. Now it is time to stick a foot in that door, dammit, and they start to plan how to go about it.

Adam Zimmerman and Alf Powis knew that they would not be overwhelmed by welcome if they were to pop up three floors to borrow a quarter for the parking meter. But these are not men who spend their time pursuing the good opinion of opponents. A very well organized and disconcertingly well funded team of determined people was almost certainly studying ways to sandbag them and storm the walls of their company, and what did they do? They went out and launched an aggressive and altogether sanguinary assault on someone else's company.

Twelve

Trouble at Noranda

The target that Alf Powis and Adam Zimmerman singled out for takeover, while their own corporate base was under siege, was a formidable one. They decided to take a swipe at MacMillan Bloedel, the largest forestry company in the country. At the time the gold-dust twins rode into Vancouver from Toronto to capture it, MacBlo had about twenty thousand employees and dominated the massive British Columbia forest products industry. The struggle took six weeks, and it was an unlovely affair. As Powis could easily comprehend, management at MacBlo was not uniformly delighted to have another company walk in, slap down the cash and tell them to move over. There was a fight. Unopposed takeovers are not too difficult. Opposed takeovers, as anyone at Brascan could have told Powis and Zimmerman, are expensive, violent dust-ups. But Powis and Zimmerman wanted MacBlo, and MacBlo could do all the kicking it liked: it was damn well coming aboard.

The bidding was fierce. Noranda had to compete with the B. C. Resources Investment Corporation, a provincial body determined to keep control of the mighty MacBlo in British Columbia. Powis and Zimmerman had to sweeten their offer, and they did. Ultimately they won over more than twice the shares they needed, and they won control of MacBlo. They took 49 per cent of the company; it cost them $626.5 million. Already you probably suspect how they paid for it.

Some cash, to be sure. But the rest they paid for in shares of Noranda, and the angry men back in the Brascan boardroom got even angrier.

The price that Alf Powis paid for MacBlo diluted the Bronfman position in Noranda even further, and for Trevor Eyton it was the last straw. At the time, one of Brascan's top people told *Maclean's* magazine about the way the Noranda people were treating them: "We pass them in the corridors, we ride with them in the same elevators, we meet them at the same cocktail parties, but every vibe we ever get from them is that they don't want to talk to us."

Brascan's board outlined several ways to proceed against Powis and Zimmerman, and finally they settled on one. The Brascan briefing paper rings with resolve:

> Having received no encouragement for a negotiated settlement or any hope of board representation, Brascan should consider increasing its position in Noranda to a level where it will be assured of a shareholder input, and thus avoid continued dilution of shareholder value. The purchase of additional shares has certain advantages: if successful, a more influential position in a world-class company is obtained; and, in any event, the present untenable situation would be resolved within a reasonably short time frame, in that Brascan would obtain either appropriate shareholder input or the opportunity to sell its investment for a substantial gain.

Alf Powis knew the fight was over. "We're looking at $2 billion mobilized against us," he told his people. "Control of Noranda is going to be bagged one way or another."

He was dead right. Brascan formed a partnership with the Caisse de Dépôt et Placement du Québec. The Caisse is a government-controlled corporation that invests the money for Quebec's provincial pension fund. It controls a staggering capital pool and was no great friend of Alf Powis and Adam Zimmerman, having received much the same treatment as Brascan when, as another large investor in Noranda, it had sought representation on the board. When they were through lining up their support, Trevor Eyton and the Caisse's Jean Campeau could virtually have walked in and written out a cheque for Noranda.

Today the Brascan position in Noranda is something like 46 per cent of the company, a colossal stake. And when there is a meeting of the board of directors, there are quite a few of them who have only to come down three floors.

The Brascan takeover of Noranda, the effective seizing of a control position in the company's affairs, was accomplished, finally, in August of 1981. What the Bronfmans thought they were getting was one of the top mining companies in the world. Here are the exact words of one of their briefing papers on the great prey they sought:

> Firstly, Noranda is favoured by being a Canadian company deriving about 80 per cent of its income in this country. Canada has the second-largest land mass in the world, much of which is geologically favourable to minerals and climatically favourable to forest products. Worldwide, there is no major natural resource company with a comparable product mix of metals, energy, and forest products. Noranda's blend of natural resources gives the company a place among the top five or six mining companies of the world. In 1980, Noranda's net income was exceeded only by Alcan, Anglo American, Amax and was larger than that of Rio Tinto.

These are heady words of praise. Rio Tinto is an almost legendary British corporation, powerful, rich, adroit, a synonym of corporate glory. But to place Noranda firmly in that pantheon was no more than the company's due.

This, then, is the company the Bronfmans and their clever platoon of strategists thought they were getting back in 1981. And of course that *is* the company they got. But it has fallen on hard times. Very hard times.

Alf Powis is sitting in the chairman's office at Noranda. That he has remained there after his determined battle to thwart the Bronfmans' takeover is a testament to his abilities, and to the fair-mindedness and downright canny appraisal of Trevor Eyton. Noranda is the love of Alf Powis's life, and no one, not even his bitterest enemies, would suggest that his devotion to the company has ever been less than religious in its intensity. Now that the fighting is over, everyone can be a model of aplomb.

"I may not like what happened," Alf Powis affirmed later, "but I'd like even less some set of rules that said they couldn't do it. I believe in the markets operating freely — though I'm not too sure I really believe in government-controlled pension funds participating in control bids. [This was a dig at the Caisse de Dépôt et Placement du Québec.] But there are lots of things I don't like that I've got to live with."

Competent chief executive officers are not exactly as thick as trees, and so it is prudence as much as magnanimity that keeps Alf Powis where he is. From his office the towers of the T-D Centre and First Canadian Place look close enough to reach out and touch. Physically it is an illusion, but metaphorically it is an image of the closeness of the Canadian corporate world. From one tower you can sit and watch the moves in the next. On the very day that we are sitting in his big office a friend of Powis's, Hartland Molson MacDougall, is moving from the top of First Canadian Place, where he was a vice-chairman of the Bank of Montreal, to the top of the Royal Trust Tower, where he will be chairman of Royal Trustco.

Everyone knows where everyone else is sitting in those towers. There is a pool of executive talent, and sometimes it seems as if they should rig bosuns' chairs from one tower to another to save the shifting executives the time and trouble of taking the elevator to the ground, walking across the street, and taking it back up.

Powis's desk is set against the west windows, his back to the glass. Kitty-corner, in the northeast of the office, a pair of gold-coloured sofas bracket a low coffee table. The room is awash in the soft northern light of autumn, and the deep green carpet glows like a new-mown lawn. Powis is wearing the court dress of the establishment: a beautifully cut suit of grey flannel, black, toe-capped laced shoes and a white shirt sufficiently uncompromising that it would have pleased Queen Victoria. His tie is a deep maroon and bears a design of rampant lions.

All of this is very proper, but it might be a more accurate reflection of the current state of Noranda if the lions were lying on their backs with their little paws in the air. For the company that the Bronfmans bought turned turtle so fast that they must have wondered what had been holding it upright in the first place. Powis is chain-smoking as he sits back in one of the sofas and talks about the troubles Noranda is in.

"We are looking at price levels for our products now — most of our products — that are the worst levels in fifty years. The year 1982 was a blood bath. Not since 1932 had we received such poor prices for our products."

Noranda's annual report for 1983 tells the sad story in figures. The giant company had revenues of more than $3 billion, *and yet it posted a loss*. Noranda lost something like $34 million in 1983. The annual report puts it like this:

> The recent recession has been particularly devastating for producers of primary materials. This has meant that, because of the relative importance of our resource industries, the impact of the recession on Canada

was more severe than was the case for most of our trading partners. This has led to a fundamental change in perceptions in many circles.

A few years ago, it was widely perceived that Canada's wealth in resources was a guarantee of future prosperity, and that companies engaged in resource production could carry very heavy tax and regulatory burdens without damage to these industries and to the economy in general. Events have proved this perception to have been totally perverse . . .

Canada's competitive position in the world at the moment is a difficult one. This is due in part to the fact that our inflation rate and cost increases have been higher than those of many of our competitors. However, a major factor has been the strength of the Canadian dollar in relation to all currencies except that of the U. S. When this was coupled with major competitive currency devaluations by competing countries in Latin America, Scandinavia and Africa, the impact was nearly lethal.

Nearly lethal!

A company like Noranda does not report to its shareholders, to every securities analyst in the country and to anyone who can read that it has been buffeted in a nearly lethal fashion unless it has seen the grim reaper pass by so close that it feels the visit had better go on record. That is why there is so much haste at a place in northwestern Ontario where a couple of years ago you would have seen nothing more imposing than a wall of trees. And that is why Noranda's corporate colour, blue, is being tacked onto the sides of a booming complex of buildings with such purpose. There is something down there that Noranda needs, and if it doesn't start coming out of the ground in a hurry and get refined into those neat little bricks that people like to keep in vaults, then there will be weeping and rending of garments on the forty-fifth floor of Commerce Court West.

"I won't say that Hemlo is the absolutely brightest star in our sky at the moment, but it's certainly *one* of the brightest stars."

There is a long, long pause as Powis runs through some of the terrible scenarios that plague a man in his position. "You know, I don't hang over the place all day long. I don't get involved in running the operation on a daily basis. But if there is any trouble up there, then I get involved. Fast. And I get heavily involved."

There has been plenty of trouble. The first contractor that Noranda had on the site went bankrupt, leaving Noranda with a lot of half-completed facilities and a hole going nowhere. There is no point in bemoaning fate. You just get on with the job. Noranda found another contractor, and now it's digging away up there like a badger after a groundhog.

And then there was the problem of the disputed claims with Lac Minerals. Powis explains:

"You know, what happens when you stake claims is that you never get it exactly right. The prospectors never take that kind of time. You realize when you're staking a claim that there's only a one-in-a-million chance that there's going to be anything on it — so you *never* stake it that carefully." Powis butts out one cigarette, pauses, thinks for a moment, then shakes his head at the memory and taps another cigarette out of the package.

"When the title was unclear," he says carefully, referring to the three claims put into dispute by the one stake erroneously placed on the Williams claims, "well . . . we were hanging out to dry with $290 million invested up there."

Driving onto the Noranda site on the Hemlo field is a fascinating sortie into a world of feverish and monumental labours. The checkpoint, one hundred yards in from the highway, is a lowered gate. The guard asks your name and calls in to verify that you have an appointment. If you are cleared, he issues you with a pass. The pass is to be surrendered when you exit. If you try to get off the Noranda camp without being able to produce a pass to show that you were there legitimately, you may expect to spend some very heated moments answering a lot of pointed questions.

The trailers occupied by Noranda Explorations, the geological wing of Noranda Inc., are just up the hill from the guarded gate, set into a grove of trees. Hundreds of thousands of lengths of core rest in labelled core racks around the site. The geologist's camp is friendly and relaxed. The clean, spacious trailers are bright inside and the geologists poring over their strange, colour-coded maps are all at ease in their comfortable quarters. After all, much of their work has already been done on The Hemlo. They are just probing the edges of the established ore body now and examining the rock that comes up out of the steadily deepening shaft.

Peter Brown, a young geologist detailed to take the visitor around the site, backs one of the high-slung, four-wheel drive trucks out of a line and bounces out of the camp back onto Noranda's main road. Just before we reach the construction site, there is a place where the road divides around a curious rock formation, a clump of rock comprising several strikingly different geological types. There is a post driven into a crevasse on top of the formation, and the sign nailed there says: Mount McKinnon.

"Don just liked this rock," Brown explains, "so we left it here."

The site simply stuns. Everywhere, men and trucks and bulldozers and cranes further the wild pursuit of the gold. The mighty headframe, clad in the pale blue steel of all the Noranda buildings on site, rises high above the dust clouds of the tumult below. Thick steel cables run from the top of the headframe to the building which houses the powerful hoist machinery. Inside the hoist building, enclosed behind glass in a silent booth, the hoist operator watches his sophisticated bank of instruments with care.

The drums that hold the cable are huge, at least twenty feet in diameter. There are five of them, for the shaft is large enough for five separate elevators down into the earth. Some of these are reserved for the ore-carrying skips, others for the cages which bear the miners up and down. The hoist operator controls his plunging charges by watching the spinning dial before him which indicates the depth of the rising or falling platform.

One of the blue-sided buildings houses the milling plant, or concentrator. This is the facility that takes crumbled rock in at one end and delivers gold bars out the other.

The first step is the reduction of the rock from big chunks of ore into smaller particles. This happens in the grinding mills, huge revolving chambers packed with steel balls where the rocks get smashed and pounded literally into dust, pulverized to the consistency of flour, which must be mixed into a slurry, or water mixture, in order to be handled with any degree of control. So fine is the material into which the rock is pounded that it must be able to pass through a screen of 200 mesh. The term 200 mesh refers to a screen so fine that every square inch is crossed by 200 wires north and south, and another 200 wires east and west. The ensuing diameter of each tiny opening will be seventy-four microns, or seventy-four one-thousandths of a millimetre. Ninety per cent of the ore must be crushed fine enough to pass through this microscopic grid.

(As fine a process as this sounds, it is by no means the finest consistency to which rock is crushed for metallurgical purposes. At some of Noranda's operations for other metals, the ore must be crushed *twice* as fine, so that it can pass through a grid of openings only thirty-seven microns wide, taking 97 per cent of the ore. Compared to such fine procedures, the flour-sized ore particles at Hemlo are coarse.)

The reason that the ore must be so finely crushed is to enable the leaching solution — the solution that is supposed to remove the metal from the ore — to get right to the metal. The leaching solution cannot do its job if the speck of metal it is supposed to grab hold of is encased in an inch of granite. Metallurgists refer to this process of grinding the ore to a fineness that releases the actual specks or grains of the sought-after metal as liberating the metal. Once the ore has been ground up so that the little bits of metal occur freely within it — are liberated — then it is ready for leaching.

A solution of sodium cyanide is introduced to the slurry. Gold has an affinity for cyanide, and so the gold changes location. At least 95 per cent of the gold moves from the ore and into the cyanide solution. The next step is to get the gold-bearing solution out of the slurry.

The slurry at this point is about half water. It is the consistency of pancake batter. To this batterlike mess are added coarse particles of carbon. This carbon is coarse only to metallurgists and in relation to the size of the particles in the rest of the slurry. The carbon chunks must still be small enough to pass through a hole one-twentieth of an inch wide. Carbon is used because it is so porous that the actual surface area available is many times the mere dimensions of the outside. For example, a piece of carbon one cubic inch in size might have a total surface area — when you add on the surfaces of all the canals and holes that penetrate it — of a square *yard*.

When the carbon is added to the slurry, the solution-borne metal tends to deposit out of the solution and onto the carbon. This is called migration, and the gold is said to migrate to the carbon. The coarser carbon is then screened out of the slurry like gravel out of sand. This achieves a concentration of gold from an ore where it constituted *0.30* ounces per ton to a carbon formation where it constitutes *200* ounces per ton of carbon.

The gold is dissolved off the carbon in solution and deposited via electrolytic cells onto sheets. From these sheets the gold is scraped, melted and cast into bars. The bars produced at the mine site will be about 80 per cent pure gold, not fine enough for the London good-delivery standard, but easily fine enough for sale. Some of the gold will be sold this way and refined later by the purchaser. Other bars will be sent by Noranda for further refining at the Canadian Mint.

On the site, super-wide pipes to carry waste are grappled in mechanical jaws and trundled off to be welded into place at the far end of a pipeline that snakes across the ground, up a hill and out of sight. There are enough concrete blocks to build a town. There are conduits to carry electrical wiring, and great ducts for heating the buildings during winter construction.

Scores of mobile homes are parked and secured together in one corner of the site. This is the camp for the men, where cooks turn out hundreds of meals three times a day, and where television and movies and pool are available. The bedrooms here — each labourer has his own room — are panelled in sheets of imitation wood. Each room holds a bunk, an easy chair, a writing table, a small bookcase and a closet. Washrooms are located at the end of each unit, one large facility serving about forty men.

There is a whole block of trailers set aside for offices. Here are the secretaries, copiers, cabinets, communications facilities, clerks, accountants, supervisors, payroll personnel: every person, pin and stapler needed to husband an investment of $290 million until it becomes a gold mine.

''Really, Hemlo couldn't have come at a worse possible time for us,'' Powis says. ''In the autumn of 1982 we'd just lost $140 million. We had to make a fundamental decision. They had discovered something up there that *looked* like it might be very important. We had to — at a time when we were pushing overall austerity — we had to decide to make a colossal investment. We went in. *Very* cautiously.

''What we thought was that we could probably get our money back — our initial investment — with only two million tons of good ore. Now, we *knew* that much was there. So even if nothing else showed up, we could recover our money. We would build only a shaft on The Hemlo and truck the ore up to our Geco mine at Manitouwadge and treat it there.

''In all, we made what was a very tight deal.''

Noranda Inc. Almost sixty thousand employees. Assets of $6 billion. Revenues of $3 billion. Up there with the biggest and the best. Fallen on hard times and looking very anxiously at a place on the Trans-Canada Highway that nobody even heard of until two guys from Timmins walked into the bush, staked it and flogged it to some very smart people in Vancouver. Noranda, speaking with the voice of its chairman:

''What's unusual is for a gold mine to have more than two, maybe three years of production ahead of it. Here we're looking at *twenty years*. And we've got those twenty years based on very little drilling. Every time we poke a hole down there, we find something.''

Pause.

''I'm not saying there's anything special about gold. But there's something special about so much of it.''

Thirteen

The Rand

There is not all that much gold in the world, even if you add it together. If you were to assemble every single brick of gold from the vaults of all the world's banks right now, you could fit the whole load quite easily onto the deck of a single modern oil tanker. The tanker might be riding a little low in the water, mind you, and you might have difficulty getting even so sanguine a firm as Lloyd's to insure the cargo. For it would be worth rather a lot.

It would be worth $2 trillion.

The reason that all this material of such overwhelming value can be assembled in so modest a space is that gold is so dense. If you were to add up every scrap of gold that has been mined from the earth and plucked out of streams since 4000 B.C., the total weight would be less than ninety thousand tons. In fact, if you could melt it all together and shape it into one great cube, the sides of the cube would measure only eighteen yards each. Every single speck of refined gold extant in the world today could fit into that eighteen-yard cube.

Most of the gold that has been mined in the world till now has been mined in the last one hundred years. For example, in the whole of the seventeenth century the world output of gold, as far as experts can determine, was 750 tons. In the last half of the nineteenth century, on the other hand, 10,000 tons came out of the ground. This wild torrent of gold began to flow into the world as the result of the great gold rushes of California, Australia and Canada. But all of this — Canada, Australia and California taken together — was like a grain of sand on the beach compared to South Africa. There is nothing on earth like the

South African goldfields. There is nothing richer. Nothing is broader or deeper. No treasure is like the treasure of the Rand.

South Africa's gold is contained in an awesome geological structure three hundred miles long. From Johannesburg, the capital of the gold industry and the country's mightiest city, the goldfields stretch to the east and west in Transvaal, and down to the southwest into Orange Free State. This is the great Witwatersrand Basin, the richest treasure-trove on the planet. There is the Far West Rand goldfield, forty miles west of Johannesburg. Ninety miles west of the city the reef nears the surface at the Klerksdorp field. From there it arcs away and down into the Orange Free State, surfacing two hundred miles away.

The most famous of the great mining tycoons who led the powerful houses which exploited the Rand was Sir Ernest Oppenheimer. It was Oppenheimer who told his architect, when he was commissioning the landmark building which has stood at 44 Main Street, Johannesburg, since 1937: "I want something between a bank and a cathedral." He got it. And he revelled in it. Oppenheimer controlled virtually single-handed South Africa's, and therefore the world's, gold and diamond industry for forty years. And with his legendary boldness, he made decisions that had a profound effect upon the gold mining business in South Africa long after.

Sir Ernest Oppenheimer died in 1957, but only a few months before his death he settled once and for all a dispute that had divided gold experts on the Rand for fifteen years. The debate centred on whether or not to start the West Deep Levels mine in the Far West Rand field. Western Deep, as mining people called the deposit, would require a heavy investment of money and technology, far heavier than the South African magnates were used to, and they were used to heavy investments.

To get at the deposits of the Western Deep, engineers would have to sink the workings down to 13,500 feet, a depth of more than two and a half miles. And when they got there, the engineers expected that they would encounter rock temperatures of 130 degrees Fahrenheit. Problems like these would have been enough to convince other men that there were easier ways to make money. But these were not the only problems. Western Deep lies in the Gastrand Hills, and the Gastrand Hills are simply running with water. So vast are the reservoirs which lie in pockets throughout the range that Oppenheimer's engineers estimated that every single shaft would have to be able to pump itself clear of thirty million gallons of water a day.

But Oppenheimer was determined. In July of 1957 he watched the first drills bite into the ground of the Orange Free State, headed for the most far-reaching mine development that South Africa had ever seen. Five years later, in 1962, the Western Deep began to produce. In the next sixty years, the mine is expected to produce two thousand tons of gold.

Oppenheimer's son, Harry, who took over the reins of Anglo American when his father died, has proven to be as masterful a businessman as his wily parent. Witness the famous dawn raid.

The dawn raid was an operation conducted against another mining house, Consolidated Gold Fields of London. Oppenheimer assembled a massive line of credit, rounded up some key blocks of shares and prepared to take a run at Consolidated. It began early one morning in February of 1980, and it ended about one hour later. Oppenheimer opened the assault on his competitor by phoning its offices in London promptly at 9 A.M. What he told them, in straightforward terms, was to get ready to take orders from a new boss, because Harry Oppenheimer was about to buy his way aboard. Gentlemanly scruples satisfied, he went to work, grabbing 25 per cent of the company before the hour was out. The purchase gave Oppenheimer's companies the largest block of shares in the London company's parent firm, Gold Fields of South Africa, and ultimately allowed Harry Oppenheimer to extend Anglo American's control to the point where the house held sway over 60 per cent of South Africa's gold production.

The capital costs associated with deep mining in South Africa are enormous. In the Transvaal, for every one hundred feet of depth, the temperature rises one degree Fahrenheit. This makes refrigeration and ventilation costs extremely high, but there is no other way to send men into the ground at such depths. At the Western Deep, for example, the air-conditioning equivalent of the vast system at work in the mines would work out to one home- or office-size air conditioner for every three cubic feet of space. Obviously such a technique for cooling the work area would be insupportably expensive, but that is the effect the engineers must achieve. There are several ingenious methods they use to achieve it.

Between shifts, the rock face is sprayed with cool water, a commodity they have plenty of in the sodden hills in which some of the mines are located. This cools off the rock for a few hours and improves air circulation through the drifts.

Another method involves the use of special jackets for the miners. These jackets have pockets that can be filled with water, then frozen. The miners strap the jackets on at the beginning of a shift, and the jackets keep them cool for the first couple of hours. When the water in the pockets has melted and warmed, the miners take the jackets off and drop them into a deep freeze, from which they take refrozen replacement jackets.

Generally, miners in South Africa work in conditions that no Canadian miner would tolerate. The stopes — tunnels into which the ore is blasted — are dreadful, narrow, tight holes. Simply reading about them is enough to induce claustrophobia. They are usually just more than a yard in diameter. Along this close-pressing passage the miner must crawl, tracing the narrow gold-bearing reef. Sometimes he must crawl down along these constricting tunnels at an angle of twenty-five or thirty degrees, head first.

The South African mining authorities are working hard to develop machinery able to trace the reef by itself, digging in along from the crosscuts and extracting the gold-bearing ore as if it were gently tugging the filling out of a sandwich. For even as narrow as the present stopes are, the South Africans still feel they are getting too much rock along with the ore.

And then there is the matter of pay.

There are just over 470,000 gold miners toiling away in the mines that spread along the great Witwatersrand Basin and into the Orange Free State. Of these, some 430,000 are blacks. According to Timothy Green, whose *New World of Gold* is a fascinating, detailed and authoritative chronicle of the industry,

> historically, it was cheaper to send ten Africans down a mine rather than invent a machine. Apart from a period immediately after the Boer War, when indentured Chinese labour was brought in from the Far East for five years, the gold mining industry had been a sponge for African labour, usually very poorly paid. For many years, the majority of the black workers came from outside South Africa itself, simply because the mines paid so little that local Africans could do much better in other industries, and it was only in the underdeveloped economies of Malawi, Botswana, and Mozambique that the mines' pittance proved attractive.

However, two factors have changed the attitudes and pay policies of the South African industrialists.

Firstly, as the whole region becomes less stable politically, the South Africans are fearful of relying on neighbouring states to supply workers for their mines. Such is the hatred for the repressive racial policies of the South African regime that mere economic self-interest can no longer be counted on to produce a flow of workers from the wretched countries that dwell beside the wealthy, white-dominated nation. Thus, if they must use cheap black labour to remain viable — and they believe they must — then the South Africans would rather it be their own. This has tended to drive wages up.

Secondly, the price of gold made such dizzying upward leaps in the 1970s that the mine owners felt able to pass along extra benefits to their workers. In real terms in that one decade, the wages of South African miners — black miners — grew by 247 per cent.

Still, the wages are desperate by the standards of Canadian hard-rock miners. The average black surface worker in South Africa earns about $100 a month, and the average pay for an underground miner is around $250 a month. According to Alf Powis, the chairman of Noranda, it is these low wages throughout the labour-intensive South African industry that keep it going at all.

"You would never even get Canadian miners to work under those conditions, no matter *what* you paid them," Powis says. "That's for starters. The whole operation would have to be run differently, the stopes dug much larger, the cooling and ventilation improved. Then there are the wages." Powis shakes his head and just laughs at the thought of the wage spread between South Africa and Canada.

"Look, in North America, what they've got wouldn't even be *ore*." In other words, the gold couldn't be extracted at a profit.

The pressure of the work force for higher wages, the fact that capital costs are expected to rise sharply rather than decrease, and the increasing necessity for the kinds of development that demand these capital costs, are all factors that militate against South Africa maintaining its position as the world leader in gold production. But great as all these pressures are, they are nothing compared to the deteriorating internal political situation that threatens to drive the country into chaos. Measuring all of these negative forces, some analysts predict that South Africa's gold production will begin to dwindle by the end of the decade, and that Russia will replace the country as the premier producer some time in the 1990s.

In fact, the South Africans themselves expect their own production to decline. Citing rising costs and poor grade as major factors, the South African Chamber of Mines has predicted that by 1987 its production will begin to drop by twenty-five tons a year. There is no country on Earth that can replace the gold lost to the world supply by a decline in South African output. There will simply be less gold to buy. And if there is less gold to buy, and people still want to buy it, then what is bad news in Johannesburg is good news in Toronto and Vancouver.

And in Hemlo.

Fourteen

The Dark Player

The Russians don't play with gold in quite the same way as the West does. In fact, the first attitude of the Communist regime was contempt. In an essay he wrote in 1921 called "The Importance of Gold Now and After the Complete Victory of Socialism," Lenin wrote that gold would be utterly useless in the perfect socialist future that awaited all of mankind when the doctrines of the Communists were established around the globe. In fact, asserted Lenin, the only thing that gold would be good for in the socialist world would be to cover the walls and the floors of public washrooms.

This has not turned out to be the case, and you will look long and hard through Moscow before you will find a public washroom covered in anything but a lot of grey tiles. Even Lenin recognized that the washroom-paving would have to wait, and that meanwhile the new Russia should "sell [gold] at the highest price, buy goods with it at the lowest price." This is recognizable advice, and you might as easily hear it on Wall Street as on Nevsky Prospekt. Lenin knew this, and admitted that a compromise was in order. He wrote: "When living among wolves, howl like wolves."

If you were looking for a motto for the Russian gold traders, that would serve about as well as anything. They are wolves, those Russians, but they feel they must be. The world of gold trading is not peopled by men and women of unvarying altruism. The world of gold trading holds more than a few cutthroats, slit-purses and footpads. And if you intend to venture out into the world of the gold traders and make a buck, well then you had better know how to do more with your stash than make jewellery.

Don't worry about the Russians. They know.

The Russians run their operations for international bullion dealing from the anonymous premises of a Zurich firm known as Wozchod Handelsbank. There is no point in going into Wozchod Handelsbank and trying to open up a chequing account. If you made it past the front door, which is very unlikely, the only thing you would run into would be a wall of blank Slavic faces with raised eyebrows. Wozchod Handelsbank is not really even a bank. It is simply a wholesaler. What it wholesales is gold. The very best, most highly refined gold on Earth. Russian gold.

There are a lot of people around who would like to know exactly how much gold the Russians have. Although they are not worried — so massive and bottomless are the riches of the Rand — the South Africans would like to know. As the world's third largest producer, the Canadians would like to know. And you can bet that there is not an intelligence agency in the West that does not spend a lot of time trying to find out to the penny what the Russians have in the way of gold. Certainly no one has come up with an authoritative figure. When the Russians want to keep their mouths shut about something, they can draw a string around their country so tight that the only thing leaking out will be Tass.

Knowing how much gold the Russians have is knowing how well prepared the Russians are to wage international diplomacy with a heavy hand. If they have a lot, then they can stand a few years of characteristically miserable Russian harvests and still afford to keep the boys in Afghanistan. If they have a lot, then they can afford to pay for all the technology they steal and bribe and generally hijack out of the West in one way or another. They need the technology; they need the wheat when their crops fail; they need the hard currency. To get hard currency they have to pay hard currency. The only hard currency the Russians produce is the stuff that comes in bricks stamped "999.9 fine".

If the Russians have a lot, then they can manipulate the price of gold, pay for terrorists, buy bombs, feed their people, hassle the West, subsidize oil to the eastern bloc, repair their crumbling industrial infrastructure, raid the U. S. dollar, rattle the stock market or conduct any of the other day-to-day business that comes along on the road to peace and harmony for all people. If they don't have a lot, then maybe some lipless capitalist — no names — can back them up against the wall and step on their toes.

Like the czars before him, Joseph Stalin recognized the value of gold. Determined to extend the reserves of Russian gold and establish new gold camps throughout the Soviet Union, Stalin turned to A. P. Serebrovsky, a trusted technocrat who ran the Russian oil industry. In 1927 Stalin made Serebrovsky the head of a new government agency, Glavzoloto, literally Gold Trust, the Soviet body that still runs the nation's gold mining efforts. Serebrovsky, under cover of being a professor from the Moscow School of Mines, travelled to the United States to take a first-hand look at how the capitalists ran their operations. He made his way to Alaska, for he wanted especially to examine gold mines which were run under the same climatic conditions to be encountered in Siberia. While he was in Alaska, the Russian gold czar met John D. Littlepage, an American mining engineer.

No sooner did Serebrovsky get back to Moscow than he huddled briefly with his political superiors, received their permission, and within a week a lucrative contract was on its way to Littlepage by diplomatic courier. The assignment: modernize the Russian gold industry. Littlepage accepted the offer, and from 1928 to 1937 the tall, rangy and powerful American roamed the goldfields of the Soviet Union, travelling by train, plane, cart, horse, mule and a battered, toiling, wheezing Model A Ford.

Littlepage transformed the Russian goldfields.

A new fleet of dredges — modern machines powered by steam and electricity — was assembled to exploit the alluvial deposits. Ultimately there were ninety of these new marvels, and they accelerated the Russian gold production tremendously. Powerhouses were built, hoists and crushers. Where the gold was mined from ore rather than scooped up as placer gold, cyanide extraction plants and smelters were erected. Littlepage convinced the authorities that he needed more skilled help, and short contracts were offered to other American engineers, working under Littlepage for the Russians.

Even a little capitalism crept into the Russian scene.

Stalin was no fool. He knew that there was nothing like greed to fire the ingenuity and determination of men. He offered rewards to the discoverers of new deposits, just as the czars had before him. A man could earn 30,000 rubles, a fantastic fortune, for locating a significant deposit. Payment was in gold, and there were special shops established in the gold mining regions where luxuries unobtainable elsewhere in Russia could be had for gold. Stalin's object in creating this curious luxury market was twofold. First, he wanted to get back some of the gold paid out as bonuses. Second, he was certain that some Russians had hidden away gold at the time of the Revolution, and he wanted to provide them with a way of spending it.

Yes, Stalin took gold wherever he could find it. But not just from his own people.

In the winter of 1936–37, the Spanish government was desperate to buy arms and aircraft for use in the civil war that was tearing its country apart. The Russians supplied the military goods, but they wanted security. The security they insisted upon comprised the main part of the national gold reserves of Spain. From Madrid to Moscow, in great secrecy, went $560 million in bullion. As the story goes, Stalin held a banquet in one of the great feasting halls of the Kremlin. Halting the proceedings late in the evening, he rose and proposed a toast to the newly arrived gold, exclaiming: "The Spaniards, comrades, will never see that gold again."

And they never have.

The goldfields of the Soviet Union are perhaps the only places in that sprawling empire where a worker can enjoy some of the advantages of the private entrepreneur. These are the independent prospectors, or *staratelу,* and it is they who tramp through the taiga and canoe along its hundred rivers searching for smaller placer deposits. When such deposits are located, the *staratelу* have two options. They can either mine the deposits alone, or form what is in essence a small corporation called an *artel*. This *artel* is compelled by law to wholesale its production to the district trust. At least one report from the town of Ust-Nera in Yakutsk province suggests that the independents can make as much as 10,000 rubles a year. In the Soviet Union, that sum would enable a man to live like a czar. Apparently the *staratelу* do just that, earning a reputation for revelry and free-spending that would fit into the California of 1850.

For most of its gold mining history, the widespread alluvial deposits of the Siberian *taiga* and tundra were the main source of Russian gold. When the sly traders of the Wozchod Handelsbank started to move their product, you could be certain that most of it had been extracted from the hostile terrain of Siberia. That has now changed.

In the barren red-sand reaches of the Kyzyl-Kum desert in the province of Uzbekistan, the Russians have established a mine which is considered to be the largest in the world. Uzbekistan is a province in Central Asia, lying between the Aral Sea and the sensitive Afghan border. The mine is called Muruntau, taking its name from a nearby mountain.

As in many Canadian mines, the gold deposits of Muruntau are contained in a lacework of quartz veins. The ore is located in a rich central shaft which plunges straight down into the ground. That is where the main ore body lies. Closer to the surface, that main shaft breaks apart and spreads out through the ground along hundreds of separate veins. To get at the tentacular structure near the surface, the Soviets are attacking the ore as if it were all in one big lump, and have simply dug an open pit to excavate the entire structure of veins. While this open pit is extracting the ore from the seams at the surface, a separate mine has been sunk to get at the richer lode below. Here the gold is wonderfully pure, almost 900 fine when it comes out of the quartz. Russian metallurgists have developed a new technique for refining this quartz-borne Muruntau gold. It is called resin-in-pulp technology, often shortened by admiring Western metallurgists to RIP. In RIP the gold migrates from the crushed ore by sticking to a resin mixture.

Like all Russian gold, Muruntau gold is further refined electrolitically to a fineness of 999.9, the hallowed Four Nines of the gold business.

Some forty thousand Soviet miners and their families live near the site in the brand-new town of Zarafshan, which the government built especially to house the mine workers. It is not an easy site to service, Muruntau. The nearest water source is eighty miles away, and the Russians have had to build a special pipeline to carry water to the mine site. But Muruntau has paid off. From that single mine the country earns more than U. S.$1.5 billion in foreign exchange every year.

How much gold do the Russians have? Certainly you will never find out by asking Glavzoloto, for you will have nothing to offer in exchange. They already know to the ounce what we produce, and the average employee of the state gold administrator could probably tell you more about the terrain around Hemlo than can the people who live in Marathon. Establishing Russian gold reserves and production figures is about as exact a science as shooting craps. But everybody likes to roll the dice, and here's what they offer:

The U. S. Central Intelligence Agency puts the annual Russian production at between 270 and 280 tons. At one time the CIA used higher estimates, but sharply revised its figures downward as the result of new information and analysis. What new information? You have as much chance of getting that as you have of getting the Kremlin to hold televised leadership debates.

Consolidated Gold Fields, the London mining conglomerate, has also taken many pains to find out what the annual Russian production is. Originally, Consolidated put it at 400 tons. Now they think this figure too high and have revised their estimate to between 280 and 350 tons. Of course, both Consolidated and the CIA could be wrong.

Another estimate is substantially higher than either Consolidated's or the CIA's guesses. At a 1979 meeting of the Joint Economic Committee of the U. S. Congress, an Oxford don named Michael Kaser presented a paper titled "Soviet Gold Production." Kaser's figure: 450 tons a year. This is a startling conclusion, at least for its long-range significance. If South African production, now at just over 600 tons a year, is declining, and if the Soviets are *increasing* their output, then fears of a reversal in the first two positions among the world's gold producers are more plausible. Russia *could* supplant South Africa by the year 2000.

The way that the Russians market their gold has immeasurably increased in sophistication. When Lenin advised his contemporaries to learn how to "howl like the wolves," they took him at his word. They are now so good at it that some of the older wolves in the forest might do well to line up for lessons.

Before 1965, all the gold that the Russians managed to scrape together from the motley operations scattered throughout their dominions was marketed by either the Moscow Narodny Bank in London or the Banque Commerciale pour l'Europe du Nord in Paris. The word *marketed* is perhaps not a good choice here. The Russians operated with all the predatory cunning of a canary. The Moscow Narodny would get an order to sell, and it sold. Never mind what the market was doing or where the price was headed. As an example, when the Cuban missile crisis was at its height in 1963, the price of gold was poised to shoot straight up. What was the Russian response to this dazzling marketing opportunity? They dumped a huge volume of gold onto the market, flattening the price in hours. Everybody had a good laugh.

It was the last one the Russians gave away for nothing.

Russian gold sales these days are run by the Vneshtorgbank — Bank for Foreign Trade — in Moscow, and a more light-footed gang of capitalist roaders you will not find anywhere else in the Soviet Union. It is the cadre of finance commissars at Vneshtorgbank who direct the operations of the playful fellows with the long knives who execute the elaborate trading strategies of the Zurich outpost, Wozchod Handelsbank. It is difficult to know what they are up to, those Russians. Experienced bullion dealers, whether from London, New York or Zurich, have long ago given up trying to estimate Russian gold production by watching the trades pulled off by Wozchod. There is simply no way to tell, so secretive are they, whether a massive bullion sale made by Wozchod represents new Soviet production moving onto the market at the direction of Vneshtorgbank, or whether the wolves have spotted a momentary spread between markets and leapt upon it, pocketing a few million dollars for a brisk two minutes' work.

The Russians have nerves of steel, and their tactics are the direct opposite of South Africa's. If the price of gold begins to tumble, the Reserve Bank of South Africa, the country's gold marketing agent, will never intervene in the market to support gold by placing a few hefty buy orders. That it *could* is beyond question. But the South Africans believe in keeping their hands off the vagaries of the free market, and they let the price of gold go where it may. The Russians, on the other hand, like to tinker, and it has rewarded them handsomely.

In one afternoon of heavy selling in 1976, the price of gold took a sharp dive from $140 an ounce to $128. The traders at Wozchod correctly identified the plunge as a momentary failure of confidence, one of those little panics that sweeps through the market from time to time. They advised their masters at the Vneshtorgbank that the price would rally if the metal was strongly supported. Vneshtorgbank gave the nod and Wozchod's traders waded into the falling market and started buying by the ton. The price steadied, levelled and, in the next few days, recovered. Within two days the Russians sold, pocketing a profit of $8 an ounce. The profit was more than a quarter of a million dollars on every ton. *In two days*.

The Russians have displayed great brilliance in understanding the thinking of the West and in using that understanding to make money. There is nothing wrong with other people believing you are a failure, the Russians will tell you, as long as you do not then *act* like a failure. Example: In May of 1978, Russia's bankers learned that the wheat crop was expected to fail. Within a few days, detailed, coded instructions from Vneshtorgbank were sent to the traders at Wozchod. Quietly, very quietly, in unspectacular amounts, the Russians began to sell gold. They knew they were going to need the foreign currency later in the year to replace the failed crop at home with wheat from abroad. They also knew that the Western traders, when they heard the news of the Russian crop failure — and they would hear — would go soft on gold, allowing the price to fall dramatically, anticipating a massive Russian dump onto the market to raise cash for the wheat purchases. But the Russians were way ahead.

By the time news of the failed harvest leaked out and was confirmed by the Russians, it was October. The Russians has been selling their gold for five months. They had all the cash they needed. Sure enough, anticipating a flood of Russian gold, the bullion price fell on all markets. The Russians' response? They just stopped selling for three months.

So unreasonable is the fear of some Westerners that they believe the Russians will arbitrarily manipulate the price of gold simply to effect havoc in the bullion market. By dumping or rapid buying, they assert, the Russians can elevate or depress the price, causing difficulties among Western producers and turmoil in the currency market.

This would be true if the Russians did it. But they don't.

Russia is a serious gold trader. Its long-range strategy, as far as the most sober analysts can determine, is to use gold to purchase foreign currency for many years into the future. They are steadfast supporters of an orderly market, and their interventions in support of gold have demonstrated this in the past. They make a profit out of these interventions, to be sure, but so what? The bullion market is not a welfare office.

As the chairman of the Wozchod Handelsbank put it, ''We have no interest in cutting down the apple tree.''

Still, it is nice to have a bushel or two of your own apples. In an orchard like Hemlo.

Fifteen

Gold and Paper

We are living in a remarkable age as far as the price of gold and interest in gold are concerned. The fluctuations in our paper currencies are so great that they are difficult to understand in any rational way. And yet even gold has behaved in a volatile way in the past few years. It is as if there is some adjustment working its way through the whole monetary system, and not until we come out on the other side will we know what everything is truly worth.

Gold has its detractors and its supporters. Between the two there exist relations about as cordial as those between NATO and the Warsaw Pact. Put another way, there is a certain want of dialogue. The two sides bellow across a gulf of disagreement so broad and so deep that they can scarcely hear one another. This is fine; they do not want to hear one another. They are the bears and the bulls.

In stock market parlance, a bear is an investor who sells a commodity in the expectation that its price will fall. More broadly, he may be one who is "bearish" about this commodity as a general rule of behaviour and opinion, and thus will not even *have* any to sell.

Bulls, on the other hand, are the optimists. If you are bullish on gold, then you believe gold has a future, and you recommend holding gold. A bull is a tout. Where a bear believes that gold is a barbaric relic, good for nothing but ornament, and ought to be banished forever from the world's monetary system, a bull believes that gold is the *only* money. There are not a lot of common points here on which to erect the edifice of compromise, and the attitudes on either side tend to reflect this "no quarter" stance. Whether you are talking to a bull or a bear, what you will hear will be the words of a prelate quoting writ. You will not be invited to discuss the merit of the position.

Gold's adherents are everywhere. You may find them in any place where men acquire enough money to worry about where to put it so that next year it will still be there. If they are true goldbugs, why then they put it in gold, counsel others to put it in gold, buy gold stocks, buy gold futures and generally do anything but entrust their wealth to the capricious and evil institution of fiat currency. Fiat currency is simply paper currency, money created by the decree of a government rather than out of some negotiable commodity, like gold. *Fiat* is the Latin for "let it be done."

"I'll tell you the only thing that fiat money is good for," one Toronto goldbug has said. "You use it to pay your taxes. Ha ha! Give them back the same worthless paper they give to you, that's what it's good for. Ha ha."

You must not be alarmed when a goldbug chuckles or guffaws or mutters to himself or twists his face into horrible caricatures of loathing when the words "fiat money" crop up. A goldbug is a person to whom truth has been freely revealed, and he is constantly affronted that an imprudent world is not prepared to receive it in turn from him. To a goldbug, only gold is money. He would define money as both a medium of exchange and a store of value. That people continue to accept paper for payment when they cannot cash it in for anything valuable, like gold, is one of the abiding mysteries of a goldbug's life. Or, as one American economist and gold-standard fancier put it, "Government is the only agency that can take a useful commodity like paper, slap some ink on it, and make it totally worthless."

A goldbug loves a horror story. He will cheerfully retail lurid accounts of the repressive laws passed by a desperate French treasury attempting to force upon a suspicious peasantry the fiat currency that was issued in the anarchic days before Napoleon. In those days it was illegal for the French to even *ask* whether a debt was to be settled in gold or in paper fiat notes. And if you actually *refused* to take the paper as payment, it was a short ride to the guillotine. This is why Napoleon, who restored the gold standard to France, is revered by goldbugs.

There are plenty of other horror stories, more recent.

When Germany entered the First World War, it had reserves of

2.092 billion marks in gold, backing paper currency of 2.904 billion marks. This is pretty good coverage in even the strictest goldbug's view. But after the war, with an exhausted treasury, the Germans were backed against the wall by the victors seeking reparation payments. They turned to the presses, and soon Germany had 250 billion marks of currency issued, with gold reserves of only 467 million marks. Here is what resulted, as described by a former editor of the newspaper *Frankfurter Zeitung:*

> Large laundry baskets filled with paper money had to be carried into the editorial conference room where the editors would sort it out, count it, and distribute the pay.
>
> As soon as everybody got his bundle, he'd rush out to buy whatever he could. Anything was more valuable than money. More and more people turned to speculation and black marketing in cloth, precious metals, foreign bills and so on. The result was that the output of industry sagged.
>
> Because of price controls and rationing, goods were scarce in cities. The railroad stations were jammed with people going out to the peasants to bargain for food.

The depradations that war makes upon an economy are a given of history. The lessons that have been learned by human beings over thousands of years, and the way people respond to the turmoil of war, is almost classical, at least in the older societies.

As the Vietnam War drew to a close, many Vietnamese — Annamese and Chinese both — could see that the United States was headed for a bitter and demoralizing defeat. The mighty, powerful and fabulously rich nation which had fuelled and driven the war up and down the lengths of the land was collapsing under a weight of confusion, self-recrimination, failure of will and appalling generalship. The Vietnamese saw this happening, knew their protector was failing fast and took the ancient precautions of their people.

There is a form of gold manufactured in Hong Kong and Singapore called a tael. A tael is a thin strip of gold which weighs about an ounce. Taels are easily stored; they can be wrapped around a limb or otherwise moulded to the body, and hidden beneath the clothing. Thus secreted, they are easily smuggled out of war zones, where tumult reigns.

Many of the Vietnamese who made it to America had these taels of gold with them, and they were sold at such U. S. refugee camps as

Fort Chaffee, Elgin Air Force Base and Indiantown Gap Military Reservation. The best estimates for the amount of money the refugees were able to take out of Vietnam in this form is about U. S.$250 million. The Vietnamese were suspicious, as they have been suspicious of all governments for a thousand years. Only U. S.$15 million was sold openly at the camps to legitimate American bullion dealers. The rest simply vanished into the pool of liquid wealth that stirs forever beneath the world of paper.

What is paper money? Paper money is money by decree. Originally, notes were promises. In fact, many of them, including Canadian currency, bore legends testifying to the fact that they were convertible into specie on demand, and that specie was usually gold. Now the notes in Canada say, "This note is legal tender." Legal tender means something that must be accepted in payment of a debt.

Thus, no one, no matter how skimpy his faith in the dollar, can refuse to accept it. It is the same in the United States, where a lot of our bad ideas come from. Here is, verbatim, a particularly telling exchange which took place in Washington in 1941. Congressman Wright Patman was questioning Marriner Eccles, governor of the Federal Reserve Board, at a meeting of the House Committee on Banking and Currency.

> *Patman:* How did you get the money to buy those $2 billion worth of government securities in 1933?
>
> *Eccles:* We created it.
>
> *Patman:* Out of what?
>
> *Eccles:* Out of the right to issue paper money.
>
> *Patman:* And there is nothing behind it, is there, except our Government's credit?
>
> *Eccles:* That is what our money system is. If there were no debts in our money system, there would be no money.

This is as good a definition of fiat money as exists anywhere. There is nothing to it. It is a promise to pay the noteholder precisely nothing. If you have confidence in the government which has issued the currency, then that currency is worth something. If you lose confidence, then the currency must be supported by artificial means, by control. It becomes a soft currency as opposed to a hard currency.

The U. S. dollar is a hard currency. Today it is enjoying a pleasant ride at the head of all world currencies, the undisputed leader. But even if it were to fall, you could still take it into a bank anywhere in the world and exchange it for the currency of another country. Inside or outside its borders, the U. S. dollar is allowed to trade against other currencies according to whatever the free market decides is equitable.

Not so with soft currencies. With a soft currency, the government that issues the notes determines what the rate of exchange for private citizens will be, vis-à-vis other currencies.

For example, if you travel to Nigeria, then you must exchange your Canadian dollars at a Nigerian bank, inside Nigeria, at a rate set by the Nigerians. That is how they make their money. They make it by running a currency monopoly at absurd prices.

Thus, you pay much more for your Nigerian currency when you buy it there than it is really worth. This gives the Nigerian government foreign exchange. Put another way, it gives them more of your money than they could have obtained by purchasing it openly and having to pay for it what the free market decided it was worth. They save; you lose. That is why you cannot exchange soft currency for the currency of your own country once you are back. A Canadian bank knows it cannot get fair exchange.

A black market comes into being.

In a country like Nigeria, people with money are always anxious to get hold of foreign exchange so they can export wealth out of the country as a hedge against bad times. It is illegal for them to buy such foreign currencies, for the government knows that all they will do is export it, draining wealth. And so if you know whom to approach, and it won't take you long to find out, then you can buy Nigerian currency, which is what you have to use there, at a much more preferential rate than the official rate of exchange. This is the black market. Another term for it is fair exchange. There is another way it works.

Say a shopkeeper wants foreign exchange. He can offer to discount his merchandise for you if you pay him in a hard currency. It is illegal, both for you and for him, but few can resist the bargain. This last is the way most foreigners, such as tourists, take advantage of black markets in other countries. You simply sidestep the government's usurious monopoly and pay what the market decides.

Goldbugs insist that all of this would be unnecessary if gold were behind paper notes. Goldbugs are right.

The goldbugs' Mecca is London. Twice a day the price of gold is fixed in London, and the London fixing is the benchmark watched by gold dealers around the world. The fixings take place at 10 A. M. and 3 P. M. London time. The fixing that North American traders await, therefore, is the 3 P. M. fixing, which begins at 10 A. M. Toronto and New York time.

The fixing takes place in the City, the financial district centred in the old City of London. The City clusters around "the Old Lady of Threadneedle Street," the Bank of England. The tangle of ancient thoroughfares beside the River Thames is home to some of the world's most prestigious banking and trading firms. These are the firms whose members cheerfully plundered the treasure of empires, and China and India would both be a lot richer today were it not for schemes hatched in the ruthless brains that made the City powerful.

To make the fixing, representatives of four venerable bullion traders make their way to the premises of N. M. Rothschild and Sons. There, they enter a room just inside the front door and take their places at a large oval table. A fifth banker, representing Rothschilds, takes his place at the head of the table. The door is closed and the negotiations to fix the bullion price begin.

Each representative has a telephone before him, which connects him to his own firm's gold room. Also before each person is a tiny Union Jack. The little staff which carries the flag is hinged, so that it can be lowered to point forward, draping the flag onto the table. As long as one of these flags is lowered before a member of the fixing cabal, then everyone else knows that he is still in conference with his own gold room and, whether his phone is at his ear or not, the price cannot be fixed.

The rule of the fixers is that every seller must be satisfied. All prices are U. S. dollars.

Let us say the Rothschilds' fixer opens the proceedings. If gold was last fixed at $400 an ounce, then he might ask, "Well, gentlemen, shall we fix the price at $400?" Each man picks up the phone before him and the negotiations begin. The price is not acceptable to every seller, however. One wants $400.10. Obviously, that price is acceptable to the sellers who would have taken the earlier $400, so all that must be done is to match up enough buyers at $400.10 to take all the gold offered. When each gold room calls back and accedes, then the price can be fixed.

Goldbugs love the story of the London Gold Pool.

The London Gold Pool was an American idea, a way to dampen the world's appetite for gold. Since the end of the Second World War, the Americans have had it all their own way, dominating the fiscal affairs of the West with their mighty and indomitable dollar. But by the end of the 1950s, the French became disturbed by signs of profligacy that led them to charge that the United States was drowning the free world in its flood of currency, freely plundering the resources of countries who conducted their currency-making with more restraint. This general suspicion percolated through the world, registering in the quarters where people who hold a lot of money dwell. Speculation in gold began to increase and the Americans, disliking developments, persuaded their European allies to join them in forming the London Gold Pool. The year was 1961.

The purpose of the Pool was to maintain an orderly market in gold. This meant different things to different people. To Americans, it meant suppressing the price of the precious metal. To others, it meant circumventing this suppression.

The official price for gold was $35 an ounce, a figure so meaningless it might as easily have been $20 an ounce. No one knew then, as no one knows now, just how the magic figure of $35 an ounce was reached, but there it stood, fluttering as bravely over American fiscal strategy as if it were the Stars and Stripes. The tactic was to depress demand when gold threatened to rise above $35 an ounce. The Pool would simply sell gold until the demand eased. When demand was slack, the Pool would buy. The price was always $35 an ounce.

This strategy was supposed to demonetize gold, sabotage its value and convince the world it was just another item of metal. It is a little like determining to hand out diamonds in downtown Winnipeg for a dollar apiece. At first people will flood to Winnipeg to buy. But when it becomes apparent that the supply will continue, at a dollar apiece, people will lose interest, and diamonds will vanish as a valuable commodity.

But as it turned out, people don't go away. People will keep on lining up for the handout until you have no more left. As a result of the entirely understandable cupidity of men and women with capital, the central banks of the United States, the United Kingdom, France, Switzerland, Belgium, Italy, the Netherlands and West Germany rapidly lost U. S.$991 million of gold reserves. For the Swiss, this probably did not represent much of a worry, since there are plenty of reasons to suspect that it was their own citizenry and bankers who were doing much of the buying. You don't mind helping the rest of the bakers subsidize bread if your family is the only one coming round to buy it.

The French were not long in recognizing the opportunity themselves. While ostensibly supporting its aims, they themselves began to buy gold from the Pool. In 1966 the United States lost $571 million in bullion to the operations of the Pool. That same year, the French *increased* their reserves by $601 million. The next year, 1967, the French made it official, and pulled out of the Pool.

The run on the Pool began in earnest.

The United States picked up 59 per cent of the operations of the Pool, a staggering load for any country to bear, even one as rich as the United States. Between November 1967 and March 1968, the United States lost a devastating $3.2 billion in the Pool.

No alliance could support so stupendous a folly. It was only a matter of time, and the last, vicious blow fell on the Gold Pool on 14 March 1968, when four hundred tons poured into the hands of private buyers. In five months the United States treasury had lost one-fifth of its gold reserves. Even as steely and determined an Administration as the one then installed in the White House could not face down further losses, and Lyndon Johnson asked the Bank of England to close the Pool.

The famous $35 an ounce was gone for good.

Epilogue

Once the foolish and arbitrary figure of $35 an ounce had fled before the relentless plundering of the Gold Pool, the price of gold took off in earnest. An astonished world watched as gold drove up through $100 an ounce, then $200, then $300. As excitement about the metal spread through the exchanges of the great trading centres, the fire caught and spread. People who had never speculated on anything more exotic than the outcome of the Stanley Cup were casting covetous eyes at the market. Up marched gold, and ever up. Newspapers carried photographs of people lined up at the doors of banks, waiting to plunk down their money for a few neatly stamped brilliant wafers. It had to end, and it did.

On 21 January 1980 gold hit U. S.$850 an ounce, twenty-four times the artifically supported price of less than twelve years before. And that was as high as it got. After months of turbulent trading, gold closed the year at $526.75. Grieving for their money were a lot of formerly innocent punters. In the classroom of the market, bad marks come straight out of your pocket.

Gold has drifted since then, risen and fallen, fluttered and soared, slipped and slid. In January of 1985 it languished around U. S.$300 an ounce. Whether you believe that gold is down for the count, or that it is about to spring again into pre-eminence, there is one thing you cannot deny, and that is that an enormously complex and sophisticated trading and speculation apparatus has grown up around the metal, and that its allure is so bright it beguiles headlines from editors whether it is well or ill. Indeed, one entire marketplace, the Vancouver Stock Exchange, exists exclusively to fuel such heady charges as the mad rush into the bush that tore the curtains from the Hemlo goldfield.

In Hemlo the miners have reached the ore. Far beneath the ground, where the temperature of the Pre-Cambrian rock is a uniform sixty-six degrees Fahrenheit, they drill, plant their charges and blast. In Moscow they hear the blast. It does not trouble them, for they have the staggering spoils of Muruntau to keep them busy. But they will pay attention, for the Canadian miners have only begun to mine the glittering lode of Hemlo.

In Johannesburg, too, they hear the blasts on the Hemlo camp. The Canadians are starting to take out their ore. This does not trouble the South Africans, for they have the Rand. Nothing is richer than the Rand. For now. But all around them the black people seethe, the country groans, and the magnates of the goldfield look at Hemlo a little wistfully. For the Canadians are taking out their ore. In peace.

In Brazil they do not even hear the blast. In Brazil the scrabbling is so wild that there is scarcely a thing to be heard on the gold camps but the sound of tearing hands. In Brazil the gold rush is a rush of the small prospector, and though the government takes its handsome slice, there does not seem to be much in the way of an opportunity for the foreign investor from, say, Zurich or London or New York. But there is plenty of opportunity on The Hemlo.

Hemlo did not arise from some effortless fiscal volition. People made Hemlo. They wrestled it into life and fed it. This chronicle has introduced only a few of these people, and we have only glimpsed their lives. But it has been a long time since Dave Bell pulled that first gold-shot core from Hole 76. Where are all these people now?

Nell Dragovan, Dave Bell and Don Moore

When Murray Pezim dealt off control of Corona to Teck Corporation in December of 1981, Moore, Dragovan and Bell were thunderstruck. Even though Pezim had found himself strapped for cash in the midst of collapsing markets, the three felt that his deal with Teck was madness. "It was desperation," Dave Bell recalled later. "It was a shame. In return for spending a million dollars, Teck got 55 per cent of Corona's interest in The Hemlo. And we'd already spent \$1.6 million ourselves."

By this time Moore was a vice-president of Corona and all three were directors. And yet Pezim hadn't even told them about the deal. They learned about if from a friend who'd seen the news on a ticker tape. According to one report, Nell Dragovan was in tears at the news, although whether they were tears of rage or grief the report did not say. Moore and Bell resigned from the Corona board in disgust and caught the first plane back to Ontario. They were going to find another gold mine, dammit, and this time they would run it themselves.

When both men were involved in the early stages of the Hemlo play, Dave Bell had been approached by John Ternowesky, a prospector from Thunder Bay. Ternowesky and two partners had staked some claims near Terrace Bay, a paper mill town 128 miles west of Hemlo. The claims surrounded an old gold showing known as the Empress mine, and Dave Bell was interested. According to Bell's broad, regional theory, the Hemlo field was located on the remnants of the slope of a long-dead volcano. The centre of this volcano would have been to the southwest of Hemlo, on the floor of the great sea that preceded Lake Superior. So Hemlo would be sitting on the northeast arc of the volcanic slope. If his theory was true, Bell reasoned, then the northwest arc of the vanished slope passed somewhere near the Terrace Bay claims. Bell decided to go and take a look.

The ground around Terrace Bay has what geologists call "a lot of relief." This means that the topography is uneven, full of ravines and steep, cliff-sided hills. On the side of one of these steep hills, in the middle of the claim belonging to the old Empress mine, Dave Bell found what is known as an adit. An adit is simply a tunnel or drift cut horizontally from a near-vertical surface. Adits are common where topography permits miners to gain access to an ore body without sinking a shaft.

Part of Dave Bell's background was five years underground as a production geologist at the Dome mine in Timmins. A production geologist's job is to interpret the geology underground, and that is why Dave Bell knew what he was looking for when he entered the old adit with a powerful electric light and started to poke around. The adit on the Empress claim dated from 1897. Later, in the 1930s, drifts had been cut leading away from the adit and farther into the ore body. Ore had been taken out.

"We found tailings on that property that graded 0.50 ounces of gold a ton," Don Moore states flatly. "Higher than anything at Hemlo."

Bell liked what he saw inside the abandoned workings, and a deal was struck with Ternowesky and his partners for thirty claims around the Empress mine. This staking was later expanded to fifty claims. Bell began to crisscross the property in a grid pattern, taking a look at the rock and whacking at the occasional projection with his pick. On the western edge of the property, Bell found some rock that looked promising and took a sample for assaying. In prospecting terms the sample kicked, which is to say it came back with the confirmation that it contained gold. In fact, it graded 0.40 ounces per ton.

''We expanded our claim block to the west and north,'' says Moore. ''Some of the claims we picked up on our own, others we optioned. The single claim with the Empress on it was held; we optioned that. It cost us some cash and some stock.''

In the spring of 1982 Micham Exploration Incorporated went public on the Vancouver Stock Exchange. (The company name is an anagram formed from the names of Don Moore's two sons, Michael and Graham.) The principal partners in the new company were the three who had been so bitter at the Teck-Corona deal: Nell Dragovan, Dave Bell and Don Moore. The partners raised money with a public stock offering, and by May 1984 had generated so much interest in their property that a Swiss group considered acquiring Micham as its exploration arm. But business developments in another part of the world forced the Swiss to withdraw, and the slump in gold prices halted further exploration on the property.

''We've spent $500,000 in drilling and other exploration,'' Don Moore claims. ''So the ground is okay in terms of the assessment requirements for another three years. But we didn't put the property together to just sit and stare at it. If money doesn't go into the ground, it sure as hell is never going to come out of it.''

But in January 1985 the price of gold was still calling the shots, and the price of gold was saying a firm ''No'' to exploration programs from one end of the country to the other.

Nell Dragovan remains in Vancouver, Dave Bell in Timmins and Don Moore in Toronto. All three are busy with other projects, but at the back of their minds sparkles an enchanting prospect. For perhaps two billion years ago, when the volcano that Dave Bell believes blasted forth the Hemlo deposit exploded, it laid down another Hemlo.

John Larche, Don McKinnon, Rocco Schiralli and Claude Bonhomme

The members of the grubstake syndicate have done well. In October of 1984 Conwest Explorations, a company with wide interests in mining, hotels, oil and gas, bought into the syndicate's holding company, Hemglo Resources, acquiring 25 per cent. The price was $6.8 million, and Conwest paid it in preferred stock. Although the Conwest deal is the first big payoff for the members of the grubstake syndicate, the sudden infusion of new wealth into their lives has not transformed their outward circumstances very much. Rocco Schiralli lived well before and he lives well now, shuttling between his Toronto condominium and his glass-and-wood palace on Peninsula Lake. Claude Bonhomme remains the same irrepressibly good-natured and accessible man he was before the deal. Bonhomme lives with his wife and three children in a $200,000 house in the Toronto suburb of North York, and while this is not exactly struggling along in straitened circumstances, it is not where the high-rollers hang their hats, either. Bonhomme has kept the same cottage on Rice Lake, east of Toronto, that he owned before, and that he describes as "just an ordinary cottage."

"You've seen me," he adds, confident that this alone explains everything. "I haven't changed."

Perhaps the busiest member of the group is Don McKinnon, a restless, questing man who seems to be driven from place to place, always seeking something that can only be found by grabbing a plane at the airport and sailing off into the blue in ceaseless pursuit. Tracked down at Bonhomme's office in Toronto in January of 1985, McKinnon delivered this breathless account: "I just got in from Ottawa. I was up there getting research for some property I'm looking into. Two packsacks full of research . . . cost me $400. I catch the seven o'clock plane out of here tomorrow morning, back to Timmins. I'll read up that research for two days, then I'm off to Vancouver."

McKinnon details all this in the hard-edged accents of the northern bush. Only men of a peculiar toughness and resilience can stand the body-breaking conditions of the remote bush. But just because a man can stand it, doesn't mean his body is not paying a heavy price, and in 1984 McKinnon was faced with a disease that attacked him precisely because of the bold recklessness he carried with him into the bush. He contracted skin cancer.

"It spread over 70 per cent of my face. The trouble was, I never wore a hat and I was out in the sun all day long. Finally I ran into an old prospector in Vancouver who gave me hell for doing nothing about it. He recommended a doctor in Vancouver."

McKinnon had been ready to enter the Princess Margaret Hospital in Toronto for surgery, but abandoned the plans and took the prospector's advice about the Vancouver doctor. In the summer of 1984 the cancer was still visible in angry red patches all over his face, but by the end of the year the salve prescribed by the physician had cleared up the painful affliction. Now, Don McKinnon wears a hat.

Most of the properties McKinnon is interested in are gold properties, flung all over the map in Ontario, Quebec and Manitoba. Gold was his business before he staked the fabulous Hemlo, and gold will remain his business. "The Hemlo strike never made any impact; I never changed my life style. Now I have security, and before there was never any real money. But I've always done the same thing. I always travelled a lot. I work on the farm in the summer and I work in the bush in the winter."

John Larche is the only member of the grubstake syndicate whose share in the Hemlo bonanza has altered, superficially, the way he lives. Simply, John Larche went out and bought himself one great big mother of a house. Gone is the modest split-level on Churchill Avenue. Come is the handsome pile full of all the bedrooms, bathrooms, dens and family rooms that a rich man could reasonably want. The house is set in a beautiful, landscaped, six-acre spread on the outskirts of Timmins. It has a swimming pool. But Larche is still very much a prospector, and so if you were to go around the back of that new house, you might find a few things that another rich man would not have parked in his backyard. Like a five-ton truck. Or a muskeg tractor. Or a backhoe.

Larche's eldest son, Paul, is the program director of a Timmins radio station. The next son, David, has followed his father into the bush and now operates as a contractor for mining companies and as an independent prospector staking his own ground. Sharing the big house with Larche are his two teenage daughters, Nicole and Lise. There is an element of sadness in all of this, for someone is missing. Larche's wife died of cancer in December of 1982.

Still every inch the prospector, Larche maintains extensive interests in gold properties all over Ontario. For years he flew into the bush in a 1947 Piper Super Cub, already eleven years old when he bought it in 1958. Now he flies into his properties in a Cessna 185, twice as large, twice as powerful, twice as fast as the old bucket that served him so faithfully before. There are advantages to being rich, certainly, but there are drawbacks, too.

"I find that I have less time to spend in the field and more time to administer my properties. I don't have the time to prospect personally. I'd rather be out in the field than behind a desk."

Frank Lang and Richard Hughes

Two of the most tireless and seasoned promoters on the Canadian mining scene, Frank Lang and Richard Hughes maintain interests or controlling positions in a dazzling string of properties. But one of the most impressive of the Vancouver promoters' ventures is an awesome tract of land in Manitoba comprising some fifteen hundred claims and held by a company with the fiscally lusty name of Kangeld Resources Limited. Kangeld began trading on the Vancouver Stock Exchange in January of 1985. Kangeld's hegemony is close to one hundred square miles of prime Manitoba bush, and if that were all there was to it, then its early trading levels of eighty or ninety cents a share would probably represent a fair price for the stock. But it is difficult to listen to Frank Lang for very long without getting the idea that if you don't start cracking some money into Kangeld soon you could get left with a lot of dust in your eyes.

"The geology resembles the geology around our Val D'Or mine, except that the [geological] structure in Manitoba is so vast. If we find one mine there, then we could find ten or twenty or thirty mines on that property. If we find one gold mine, then there's a good chance we'll find more. And of course, unlike Hemlo, we own the whole thing. We could process ore from all the mines in one central mill with a capacity of, say, twenty thousand tons a day."

Frank Lang never stops promoting.

Alf Powis

In January of 1985 Alf Powis was still solidly in place in the big office in the northwest corner of the forty-fifth floor of Commerce Court West in Toronto. That is where he will probably stay. For the people who run Brascan approve of the Noranda management, and Brascan owns 70 per cent of a certain company called Brascade Resources Incorporated. And Brascade Resources owns, as of early 1985, 46 per cent of Noranda. Brascade was formed specifically to hold Noranda after the little spot of ugliness that smudged the first days of the, well, *relationship* between Noranda and Brascan. But since the din of battle has faded away and everyone is being gentlemanly, well then Brascade can get on with Brascan's business, which is to build up the stake in Noranda to 50 per cent, plus one voting share. Put another way, Brascan will be running Noranda.

''Noranda has gone through some very difficult times,'' a senior officer of Brascan said early in 1985. ''Prices are beyond their control,'' he added, referring to the slump that has flattened the performance of the natural resources industry worldwide. ''But we are happy with management. We were so pleased when we saw Noranda acquire that investment [Hemlo]. We couldn't praise them enough. It was wonderful work. They did a superb job and we are very, very, very pleased.''

It certainly has a wonderful healing effect, that gold mine.

By January of 1985 Noranda's main shaft had dropped past the 2100-foot level and the underground development group had struck the ore body. The assay and the thickness established by surface drilling were confirmed, and everyone heaved a mighty sigh. Actually digging down to it and hauling out the tape measure is the only way you can ever know for sure what is beneath the ground. The plans in early 1985 called for gold production to begin in March or April, with a thousand tons of ore a day going through the mill. The Golden Giant mine will employ two hundred men, all housed with their families in new accommodation built expressly for the Hemlo production workers in the town of Manitouwadge.

The Pez

Trouble came to Murray Pezim in October 1984, and was no stranger. There have been a lot of blows landed on Murray Pezim since the colourful entrepreneur first stormed onto the street back in the 1950s. But few can have been so bitter as the blows which fell during that depressing month.

Pezamerica Resources Corporation was the holding company through which The Pez held onto 62 per cent of International Corona Resources Limited, and through Corona, 45 per cent of the Teck-Corona action on the Hemlo camp. But in October of 1984 things began to sour for Pezamerica. The holding company's shares, which had once traded as high as $8, fell to $3.25. According to one report, The Pez sold 825,000 shares of his own stock in Pezamerica, and gave up eight thousand of the twenty thousand square feet of office space he rented in the Vancouver Stock Exchange tower. Even Stan, the thick-fingered, light-footed and altogether menacing presence of the fifteenth floor — Stan the cook, Stan the bodyguard, Stan the masseur — even Stan was cut from the dwindling staff. Pezzaz Productions

Inc., the company that was making and selling Greetings from the Stars, laid off some of its employees. But Murray Pezim's hold on Pezamerica was crumbling, and finally, despite the last-ditch austerity measures, he had to sell.

On 22 October 1984 *The Financial Times of Canada* reported that:

> Royex Gold Mining Corp., a Toronto subsidiary of Campbell Resources Inc., had acquired 1.6 million shares of Pezamerica and the rights to 1.2 million more. The two blocks amount to about 20 per cent of Pezamerica's outstanding shares.
>
> The price was not disclosed, but the sale gave Royex effective control of Pezamerica and stripped one of Canada's most controversial and colourful promoters of his greatest asset — a major piece of the Hemlo gold camp in northern Ontario.

Suffering from extreme depression, Murray Pezim left Vancouver and went to ground in Scottsdale, Arizona, where his wife maintains a permanent residence. The bad news was not over. Despite an investment of $3.8 million by Pezamerica in the Greetings from the Stars project, the scheme failed, and on 23 November 1984 Pezzaz Productions went into bankruptcy.

Late in 1984 Murray Pezim called Toronto from Scottsdale. During the course of the ensuing conversation, Pezim insisted that he had not wholly lost his interest in The Hemlo, and had been retained on the board of both International Corona and Pezamerica. He sounded confident and optimistic, and said that he would in time be re-entering the Canadian business world.

''When we're ready, you'll hear from us. Okay, buddy?''

On 18 January 1984 Royex issued an information circular in Toronto confirming that Pezim remained as chairman of both Pezamerica and Corona, for a combined salary of $150,000 a year. The circular outlined the intricate steps whereby Royex would swallow Pezamerica and ''create a major North American gold mining company with gold mines of various sizes and in different geographic locations and gold properties in varying stages of development. The business combination [resulting from the Royex-Pezamerica union] will result in a corporate organization having an interest in four developing gold mines and eight gold properties approaching development with the combined potential to produce 400,000 ounces annually within seven years.''

Well, great, but where does this leave the guy with the heartful of faith who brought in possibly the fattest gold camp in the Americas? Where does this leave The Pez, who hung in at Hemlo when everyone was telling him to pack up his drills, for heaven's sake, and give the place back to the moose?

On 20 January 1985 The Pez flew back into Vancouver. Without so much as a pause for breath, he installed himself in his offices in the stock exchange tower and started to pull himself out of the hole. The Pez was sounding a little forlorn, to be sure, but he was sounding a long, long way from being beaten.

"You know what happened? Gold went down, and I just owned too much stock. I had to sell some. It happens to everyone, not just me. And they're excellent people [at Royex]. I'm still on the board with them. I still have about 400,000 shares of Pezamerica.

"Slowly we're putting together a team. We suffered, but the whole market suffered. We still believe in gold. I think the turn is going to come around May."

And so the offices may not be thronging with as dense a clutch of adherents as once they held, but The Pez is getting along, thank you. Yes, there are some staff missing, but everyone in the market in Vancouver has had to cut back a bit, and The Pez is not dwelling upon the absence of the patter of little feet. No, The Pez is ready to tell you, if you'll stand still for a sec, that the American dollar can't keep this up much longer. It has to collapse. That is why The Pez and his slimmed-down team are keeping a very sharp eye out for promising ground. Because, buddy, the turn is going to come. And The Pez is going to round that turn. Bet on it.

Golden Words: A Glossary

Acid Test Testing for the purity of gold by exposing the metal to an acid or combination of acids. The tester watches for a colour change.

Alloy Gold can be alloyed, or mixed, with copper, nickel, silver or zinc. This makes the metal harder and changes its colour.

Arbitrage The buying and selling of a commodity on different markets at the same time. An arbitrageur watches the price spread between markets. He may buy in New York and sell in Winnipeg, at exactly the same time, pocketing a few cents an ounce profit. An arbitrageur must deal in extremely large amounts of gold to make any real money.

Assay A test to determine purity.

Bullion Negotiable gold. Usually, bullion must be at least .995 fine. It is available in bars, ingots and wafers.

Cage The elevator that takes miners down the mine shaft to the lower levels.

Claim A tract of land — in Ontario about forty acres — staked by a prospector or miner for mineral exploration or mining.

Coin gold Coarser than bullion, usually alloyed with silver or copper, or both, to improve toughness. Canadian coins are purer than Europeans'.

Dealing off A promoter deals off an interest in a property to another, usually larger promoter, or to a major mining company, for consideration, either cash or stock. In effect, dealing off a property is selling it.

Drift Any horizontal, or crosscut, tunnel in a gold mine.

Ductility The capability of being drawn into wire. Gold is the most ductile metal in the world, and one ounce can be drawn into a wire fifty miles long.

Electrolytic gold A refining distinction. All Russian gold, for example, is refined electrolytically, using a process that involves an electric current. The Russians refine their bars to 999 fine, effectively pure gold.

Fine gold Pure gold.

Fineness A measurement of the amount of pure gold in a bar, expressed in thousandths. Thus, a bar of 995, which dealers refer to as "two nines five," would have 995 parts of pure gold for every five parts of alloy.

Fine weight The amount of pure gold in a coin or bar.

Fool's gold Iron pyrite. Little flecks of pyrite glinting from sand and rocks are often mistaken by the untrained eye for gold. Once you have seen occurrences of real gold, however, you will never make the mistake again.

Four nines The finest bullion anywhere, 999.9 fine. Gold traders call this "four nines fine gold."

Gilding Coating with a thin layer of gold.

Goldbug A devotee of gold, a zealot, a true believer. For the goldbug, *only* gold is real money.

Gold camp The scene of gold production, or of full-scale development leading up to gold production. No longer merely a "play." The whole town of Red Lake, Ontario, for example, would be referred to as a gold camp.

Gold leaf Thinly beaten gold used for decorative purposes. The art of making gold leaf dates back to the early Egyptians.

Gold play The area of, and the promotional activity surrounding, a specific scene of gold exploration.

Gold standard A system of currency in which paper notes may be exchanged for gold.

Grubstake In the early days, a grubstaker provided the prospector with grub, or food. Today the grubstaker provides cash, in exchange for which he earns an interest in whatever claims the prospector wants to stake. Thus, a grubstake is a simple form of risk capital provided to fund a project.

Hallmark On a bullion bar, a stamped mark giving the name of the producer, or refiner, and the fineness.

The Hemlo A colloquialism for the whole Hemlo area. Loosely, when locals refer to The Hemlo, they are referring to the whole gold camp and surrounding explorations.

High-grading The theft of gold from a gold mine by a miner. The ways to smuggle gold past the security apparatus of a mine are limited only by the ingenuity of miners, and that is not much of a limitation. Every now and then the high-grading gets to be too much for the mine owners to bear, and they crack down with a will.

High-grading is one of the few crimes where the presumption of innocence does not apply. Section 583 of the Criminal Code of Canada reads: "In any proceeding in respect of ores or minerals, the possession, contrary to any law in that behalf, of smelted gold or silver, gold-bearing quartz, or unsmelted or unmanufactured gold or silver, by an operator, workman or labourer actively engaged in or on a mine, is, in the absence of any evidence to the contrary, proof that the gold, silver or quartz was stolen by him."

Karat The unit used to measure the amount of pure gold in an alloy. A karat is 1/24 part; hence pure gold is 24-karat gold.

London Delivery Bar This is the basic, world-standard trading unit of the London market. Also called a *good delivery bar,* it should weigh about 400 ounces troy, be at least 995 fine, and be marked by an assayer acceptable to the London dealers.

London market The market for gold created by the venerable institutions of N. M. Rothschild & Sons Ltd., Johnson Matthey Bankers Ltd., Mocatta & Goldsmid Ltd., Samuel Montagu & Co. Ltd. and Sharps, Pixley Ltd.

Malleability The degree to which a metal can be shaped or formed. Gold is highly malleable and can be hammered to a thinness of five one-millionths of an inch.

Melter Another term for refiner. Refiners buy mine production and refine it to London delivery standard.

Nugget Freely occurring gold, usually washed out of rock and deposited in the beds of rivers, creeks and streams. The biggest nugget ever found — in Australia in 1872 — weighed two hundred pounds.

Option The right to earn an interest in or acquire outright a block of claims through the performance of certain stipulated terms. Thus, Noranda's option on The Golden Giant enabled it to earn an interest in the property by bringing in a mine.

Placer gold Alluvial deposits of gold, usually in riverbeds where the mineral has been eroded away from the host rock. Centuries of water erosion or the pressures of glaciation free placer gold. A nugget is an example of placer gold; the flakes panned out of gravel another.

Play *See* gold play.

Solid gold Beware of this term! In a 1976 ruling, the U. S. Federal Trade Commission stated that "any article that does not have a hollow centre and has a fineness of ten karats" may be called "solid gold."

Skip An elevator used solely for the hoisting of minerals from the deep levels of a mine.

Spot price The price that gold fetches right now. Most transactions in the gold market are "at spot." In other words, you have to pay immediately. The seller must deliver the gold within two business days, and the buyer pays for it when he gets it.

Strike The horizontal direction of a mineralized zone. Thus, if an ore body strikes to the east, that is the direction in which the occurrence wanders through the host rock.

Tael Chinese gold weight. A tael is precisely 1.2034 ounces troy. Taels used to be found exclusively in the Far East, but after the Vietnam War, many began to appear in the United States, smuggled in by refugees. Taels come in many forms, some of them tailored for concealment. For example, a tael can be cut into the shape of a shoe sole and concealed between layers of shoe leather.

Gold from the Far East has lately been rather good quality. Melters' hallmarks to look for are Kim Fook, Lee Cheon and Kim Thanh.

Trenching Scraping the overburden off a long strip of rock to reveal the geological contours and formations beneath, and determine how they change over a given length.

Troy Troy is the common weight measurement system for precious metals. As well as gold, silver, platinum and palladium are all expressed in troy weights. One ounce troy is 1.09711 ounces, or 31.103 grams.

the HEMLO gold area

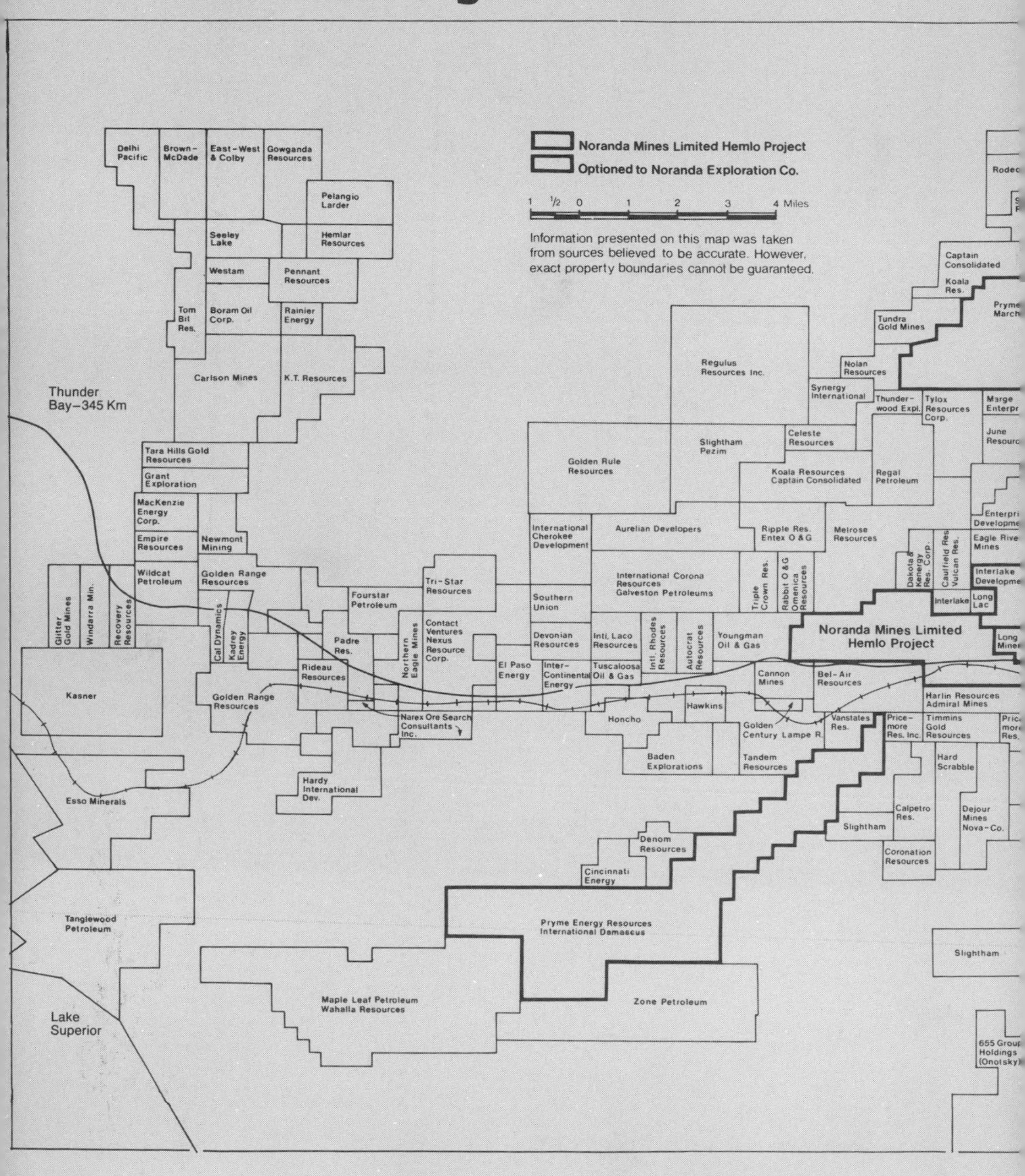